PERIODIC TABLE
OF THE **ELEMENTS**

Nonmetals

PERIODIC TABLE OF THE ELEMENTS

Nonmetals

**Monica Halka, Ph.D., and
Brian Nordstrom, Ed.D.**

Facts On File
An imprint of Infobase Publishing

Facts On File, Inc.
An imprint of Infobase Publishing
132 West 31st Street
New York NY 10001

Library of Congress Cataloging-in-Publication Data
Halka, Monica.
  Nonmetals / Monica Halka and Brian Nordstrom.
      p. cm. — (Periodic table of the elements)
  Includes bibliographical references and index.
  ISBN 978-0-8160-7367-2
  1. Nonmetals. 2. Periodic law. I. Nordstrom, Brian. II. Title.
  QD161.H35 2010
  546'.7—dc22   2009018453

Facts On File books are available at special discounts when purchased in bulk quantities for businesses, associations, institutions, or sales promotions. Please call our Special Sales Department in New York at (212) 967-8800 or (800) 322-8755.

You can find Facts On File on the World Wide Web at http://www.factsonfile.com

Excerpts included herewith have been reprinted by permission of the copyright holders; the author has made every effort to contact copyright holders. The publishers will be glad to rectify, in future editions, any errors or omissions brought to their notice.

Text design by Erik Lindstrom
Composition by Facts On File
Illustrations by Dale Williams
Photo research by Tobi Zausner, Ph.D.
Cover printed by Bang Printing, Brainerd, Minn.
Book printed and bound by Bang Printing, Brainerd, Minn.
Date printed: May 2010
Printed in the United States of America

10 9 8 7 6 5 4 3 2 1

This book is printed on acid-free paper.

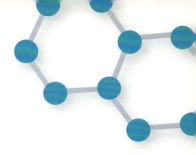

Contents

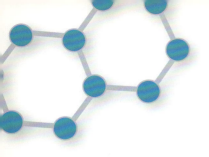

Preface

Speculations about the nature of matter date back to ancient Greek philosophers like Thales, who lived in the sixth century B.C.E., and Democritus, who lived in the fifth century B.C.E., and to whom we credit the first theory of *atoms*. It has taken two and a half millennia for natural philosophers and, more recently, for chemists and physicists to arrive at a modern understanding of the nature of *elements* and *compounds*. By the 19th century, chemists such as John Dalton of England had learned to define elements as pure substances that contain only one kind of atom. It took scientists like the British physicists Joseph John Thomson and Ernest Rutherford in the early years of the 20th century, however, to demonstrate what atoms are—entities composed of even smaller and more elementary particles called *protons, neutrons,* and *electrons*. These particles give atoms their properties and, in turn, give elements their physical and chemical properties.

After Dalton, there were several attempts throughout Western Europe to organize the known elements into a conceptual framework that would account for the similar properties that related groups of elements exhibit and for trends in properties that correlate with increases in atomic weights. The most successful *periodic table* of the elements was designed in 1869 by a Russian chemist, Dmitri Mendeleev. Mendeleev's method of organizing the elements into columns grouping elements with similar chemical and physical properties proved to be so practical that his table is still essentially the only one in use today.

While there are many excellent works written about the periodic table (which are listed in the section on further resources), recent scientific investigation has uncovered much that was previously unknown about nearly every element. The Periodic Table of the Elements, a six-volume set, is intended not only to explain how the elements were discovered and what their most prominent chemical and physical properties are, but also to inform the reader of new discoveries and uses in fields ranging from astrophysics to material science. Students, teachers, and the general public seldom have the opportunity to keep abreast of these new developments, as journal articles for the nonspecialist are hard to find. This work attempts to communicate new scientific findings simply and clearly, in language accessible to readers with little or no formal background in chemistry or physics. It should, however, also appeal to scientists who wish to update their understanding of the natural elements.

Each volume highlights a group of related elements as they appear in the periodic table. For each element, the set provides information regarding:

- the discovery and naming of the element, including its role in history, and some (though not all) of the important scientists involved;
- the basics of the element, including such properties as its atomic number, atomic mass, electronic configuration, melting and boiling temperatures, abundances (when known), and important isotopes;
- the chemistry of the element;
- new developments and dilemmas regarding current understanding; and
- past, present, and possible future uses of the element in science and technology.

Some topics, while important to many elements, do not apply to all. Though nearly all elements are known to have originated in stars or stellar explosions, little information is available for some. Some others that have been synthesized by scientists on Earth have not been observed in stellar spectra. If significant astrophysical nucleosynthesis research exists, it is presented as a separate section. The similar situation applies for geophysical research.

Special topic sections describe applications for two or more closely associated elements. Sidebars mainly refer to new developments of special interest. Further resources for the reader appear at the end of the book, with specific listings pertaining to each chapter, as well as a listing of some more general resources.

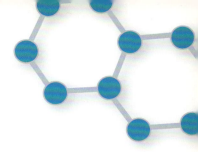

Acknowledgments

First and foremost, I thank my parents, who convinced me that I was capable of achieving any goal. In graduate school, my thesis adviser, Dr. Howard Bryant, influenced my way of thinking about science more than anyone else. Howard taught me that learning requires having the humility to doubt your understanding and that it is important for a physicist to be able to explain her work to anyone. I have always admired the ability to communicate scientific ideas to nonscientists and wish to express my appreciation for conversations with National Public Radio science correspondent Joe Palca, whose clarity of style I attempt to emulate in this work. I also thank Dr. Nick Hud of Georgia Tech and Mark Ball, aquarist at the Scripps Institution of Oceanography, for enlightening discussions.

—*Monica Halka*

In 1967, I entered the University of California at Berkeley. Several professors, including John Phillips, George Trilling, Robert Brown, Samuel Markowitz, and A. Starker Leopold, made significant and lasting impressions. I owe an especial debt of gratitude to Harold Johnston, who was my graduate research adviser in the field of atmospheric chemistry. I have known personally many of the scientists mentioned in the Periodic Table of the Elements set: For example, I studied under Neil Bartlett, Kenneth Street, Jr., and physics Nobel laureate Emilio Segrè. I especially cherish having known chemistry Nobel laureate Glenn

Seaborg. I also acknowledge my past and present colleagues at California State University; Northern Arizona University; and Embry-Riddle Aeronautical University, Prescott, Arizona, without whom my career in education would not have been as enjoyable.

—*Brian Nordstrom*

Both authors thank Jodie Rhodes and Frank Darmstadt for their encouragement, patience, and understanding.

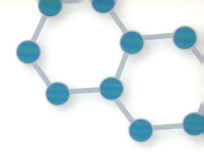

Introduction

Materials that are poor conductors of electricity are generally considered nonmetals. One important use of nonmetals, in fact, is the capability to insulate against current flow. Earth's atmosphere is composed of nonmetallic elements, but lightning can break down the electron bonds and allow huge voltages to make their way to the ground. Water in its pure form is nonmetallic, though it almost always contains impurities called electrolytes that allow for an electric field.

While scientists categorize the chemical elements as nonmetals, metals, and metalloids largely based on the elements' abilities to conduct electricity at normal temperatures and pressures, there are other distinctions taken into account when classifying the elements in the periodic table. The noble gases, for example, are nonmetals, but have such special properties that they are given their own classification. The same is true for the halogens. When referring to the periodic table, the nonmetal classification is given to hydrogen, carbon, nitrogen, phosphorus, oxygen, sulfur, and selenium. All these elements, except hydrogen, appear on the right side of the periodic table (see "The Nonmetals Corner," shown below). Hydrogen's place is at the upper left, strictly because of its electron configuration, though it has been shifted in the following table for ease of grouping.

The goal of *Nonmetals* is to present the current scientific understanding of the physics, chemistry, and geology of the nonmetals, including how the nonmetals are synthesized in the universe, when and

THE NONMETALS CORNER

				H	He
B	**C**	**N**	**O**	F	Ne
(Al)	*Si*	**P**	**S**	Cl	Ar
(Ga)	*Ge*	*As*	**Se**	Br	Kr
(In)	(Sn)	*Sb*	*Te*	I	Xe
(Tl)	(Pb)	(Bi)	*Po*	At	Rn

Bold = nonmetals; italics = metalloids; parentheses = metals; halogens = F, Cl, Br, I, At; noble gases = He, Ne, Ar, Kr, Xe, Rn

how they were discovered, and where they are found on Earth. It also details how nonmetals are used by humans and the resulting benefits and challenges to society, health, and the environment.

The first chapter is arguably the most important: Without an understanding of the simplest and most abundant element in the universe, one cannot understand the more complicated ones. Hydrogen was the only nonmetal synthesized in the big bang and was crucial in the formulation of quantum theory. The future of hydrogen research is also rich. Fusion of its heavy isotopes is considered the most likely process to result in a fusion reactor suitable for electricity production, and hydrogen fuel cells may turn out to be of great importance in alternative energy vehicles.

The second chapter discusses the element without which life would not exist—carbon. From its formation in stars to its importance in carbon dating and petroleum deposits, this chapter explains the science behind several important contemporary issues, including carbon emissions, peak oil, and climate change.

Chapters 3 and 5 discuss the gases that humans, animals, and plants need for respiration—nitrogen and oxygen—and the effects an excess or depletion of either can have when considering such diverse subjects as scuba diving, oxygen bars, automobile tires, explosives, and global warming.

The fourth chapter explores not only the astrophysics and discovery of phosphorus on Earth, but the important chemistry of this element. All plants rely on phosphorus as a building block to produce glucose, the food that fuels growth of leaves, flowers, fruits, and seeds via the process known as photosynthesis. Phosphates are, therefore, essential in fertilizers, but their presence in ponds and lakes can cause serious environmental problems.

The last two chapters cover the history and usefulness of sulfur and selenium. For centuries, humans have enjoyed the uses of sulfur from firestarters to food preservation. Selenium, on the other hand, has only recently found its niche in technology.

As an important introductory tool, the reader should note the following general properties of nonmetals:

1. The atoms of nonmetals tend to be smaller than those of metals. Several of the other properties of nonmetals result from their atomic sizes.
2. Nonmetals exhibit very low electrical conductivities. The low—or nonexistent—electrical conductivity is the most important property that distinguishes nonmetals from metals.
3. Nonmetals have high electronegativities. This means that the atoms of nonmetals have a strong tendency to attract more electrons than what they would normally have.
4. Nonmetals have high electron affinities. This means that the atoms of nonmetals have a strong tendency to hold on to the electrons they already have. In contrast, metals rather easily give up one or more electrons to nonmetals; metals, therefore, easily form positively charged ions, and metals readily conduct electricity.
5. Under normal conditions of temperature and pressure, some nonmetals are found as gases, some are found as solids, and one is found as a liquid. In contrast, with the exception of mercury, all metals are solids at room temperature. The fact that so many nonmetals exist as liquids or gases means that nonmetals generally have relatively low melting and boiling points under normal atmospheric conditions.

6. In their solid state, nonmetals tend to be brittle. Therefore, they lack the malleability and ductility exhibited by metals.

The following is a list of the general chemical properties of nonmetals:

1. Whereas very few metals can be found in nature as the pure elements, most of the nonmetals exist in nature as the pure elements.
2. Nonmetals form simple negative ions. These ions easily form ionic compounds with metallic elements. Examples of compounds containing simple ions are LiH, Fe_2O_3, Na_3N, CuS, K_2Se, and Ca_3P_2.
3. Atoms of different nonmetallic elements can form polyatomic, or complex, negative ions. Examples of compounds containing complex ions are $CaCO_3$, K_2SO_4, Na_3PO_4, and $Fe(NO_3)_2$.
4. Nonmetallic elements form covalent chemical bonds with other nonmetallic elements. Consequently, compounds of nonmetals often exist as small molecules, for example, H_2O, NH_3, and CH_4.
5. Nonmetals can exist in both positive and negative *oxidation states*. This means, for example, that nonmetallic elements tend to readily form compounds with both hydrogen and oxygen. Examples of such compounds are CO_2, CH_4, NO_2, and NH_3.
6. The oxides of nonmetals tend to be acidic when dissolved in water.

In terms of general chemical reactivity, however, nonmetals exhibit a wide range of tendencies to combine with other elements.

Overall, *Nonmetals* provides the reader, whether student or scientist, with an up-to-date understanding regarding each of the nonmetals—where they came from, how they fit into our current technological society, and where they may lead us.

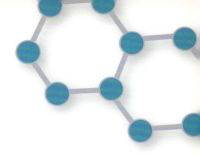

Overview: Chemistry and Physics Background

What *is* an element? To the ancient Greeks, everything on Earth was made from only four elements—earth, air, fire, and water. Celestial bodies—the Sun, Moon, planets, and stars—were made of a fifth element: ether. Only gradually did the concept of an element become more specific.

An important observation about nature was that substances can change into other substances. For example, wood burns, producing heat, light, and smoke and leaving ash. Pure metals like gold, copper, silver, iron, and lead can be smelted from their ores. Grape juice can be fermented to make wine and barley fermented to make beer. Food can be cooked; food can also putrefy. The baking of clay converts it into bricks and pottery. These changes are all examples of chemical reactions. Alchemists' careful observations of many chemical reactions greatly helped them to clarify the differences between the most elementary substances ("elements") and combinations of elementary substances ("compounds" or "mixtures").

Elements came to be recognized as simple substances that cannot be decomposed into other even simpler substances by chemical reactions. Some of the elements that had been identified by the Middle Ages are easily recognized in the periodic table because they still have chemical symbols that come from their Latin names. These elements are listed below.

Modern atomic theory began with the work of the English chemist John Dalton in the first decade of the 19th century. As the concept of the atomic composition of matter developed, chemists began to define elements as simple substances that contain only one kind of atom. Because scientists in the 19th century lacked any experimental apparatus capable of probing the structure of atoms, the 19th-century model of the atom was rather simple. Atoms were thought of as small spheres of uniform density; atoms of different elements differed only in their masses. Despite the simplicity of this model of the atom, it was a great step forward in our understanding of the nature of matter. Elements could be defined as simple substances containing only one kind of atom. Compounds are simple substances that contain more than one kind of atom. Because atoms have definite masses, and only whole numbers of atoms can combine to make molecules, the different elements that make up compounds are found in definite proportions by mass. (For example, a molecule of water contains one oxygen atom and two hydrogen atoms, or a mass ratio of oxygen-to-hydrogen of about 8:1.) Since atoms are neither created nor destroyed during ordinary chemical reactions ("ordinary" meaning in contrast to "nuclear" reactions), what happens

ELEMENTS KNOWN TO ANCIENT PEOPLE

Iron: Fe ("ferrum")	Copper: Cu ("cuprum")
Silver: Ag ("argentum")	Gold: Au ("aurum")
Lead: Pb ("plumbum")	Tin: Sn ("stannum")
Antimony: Sb ("stibium")	Mercury: Hg ("hydrargyrum")
*Sodium: Na ("natrium")	*Potassium: K ("kalium")
Sulfur: S ("sulfur")	

Note: *Sodium and potassium were not isolated as pure elements until the early 1800s, but some of their salts were known to ancient people.

EXAMPLES OF ELEMENTS, COMPOUNDS, AND MIXTURES

ELEMENTS	COMPOUNDS	MIXTURES
Hydrogen	Water	Salt water
Oxygen	Carbon dioxide	Air
Carbon	Propane	Natural gas
Sodium	Table salt	Salt and pepper
Iron	Hemoglobin	Blood
Silicon	Silicon dioxide	Sand

in chemical reactions is that atoms are rearranged into combinations that differ from the original reactants, but in doing so, the total mass is conserved. Mixtures are combinations of elements that are not in definite proportions. (In salt water, for example, the salt could be 3 percent by mass, or 5 percent by mass, or many other possibilities; regardless of the percentage of salt, it would still be called "salt water.") Chemical reactions are not required to separate the components of mixtures; the components of mixtures can be separated by physical processes such as distillation, evaporation, or precipitation. Examples of elements, compounds, and mixtures are listed in the table above.

The definition of an element became more precise at the dawn of the 20th century with the discovery of the proton. We now know that an atom has a small center called the "nucleus." In the nucleus are one or more protons, positively charged particles, the number of which determine an atom's identity. The number of protons an atom has is referred to as its "atomic number." Hydrogen, the lightest element, has an atomic number of 1, which means each of its atoms contains a single proton. The next element, helium, has an atomic number of 2, which means each of its atoms contain two protons. Lithium has an atomic number of 3, so its atoms have three protons, and so forth, all the way through the periodic table. Atomic nuclei also contain neutrons, but atoms of the same element can have different numbers of neutrons; we call atoms of the same element with different number of neutrons "isotopes."

There are roughly 92 naturally occurring elements—hydrogen through uranium. Of those 92, two elements, technetium (element 43) and promethium (element 61), may once have occurred naturally on Earth, but the atoms that originally occurred on Earth have decayed away, and those two elements are now produced artificially in nuclear reactors. In fact, technetium is produced in significant quantities because of its daily use by hospitals in nuclear medicine. Some of the other first 92 elements—polonium, astatine, and francium, for example—are so radioactive that they exist in only tiny amounts. All of the elements with atomic numbers greater than 92—the so-called transuranium elements—are all produced artificially in nuclear reactors or particle accelerators. As of the writing of this book, the discoveries of the elements through number 118 (with the exception of number 117) have all been reported. The discoveries of elements with atomic numbers greater than 112 have not yet been confirmed, so those elements have not yet been named.

When the Russian chemist Dmitri Mendeleev (1834–1907) developed his version of the periodic table in 1869, he arranged the elements known at that time in order of *atomic mass* or *atomic weight* so that they fell into columns called *groups* or *families* consisting of elements with similar chemical and physical properties. By doing so, the rows exhibit periodic trends in properties going from left to right across the table, hence the reference to rows as *periods* and name "periodic table."

Mendeleev's table was not the first periodic table, nor was Mendeleev the first person to notice *triads* or other groupings of elements with similar properties. What made Mendeleev's table successful and the one we use today are two innovative features. In the 1860s, the concept of *atomic number* had not yet been developed, only the concept of atomic mass. Elements were always listed in order of their atomic masses, beginning with the lightest element, hydrogen, and ending with the heaviest element known at that time, uranium. Gallium and germanium, however, had not yet been discovered. Therefore, if one were listing the known elements in order of atomic mass, arsenic would follow zinc, but that would place arsenic between aluminum and indium.

Russian chemist Dmitri Mendeleev created the periodic table of the elements. (*Scala/Art Resource*)

That does not make sense because arsenic's properties are much more like those of phosphorus and antimony, not like those of aluminum and indium.

Mendeleev's Periodic Table (1871)

Group Period	I	II	III	IV	V	VI	VII	VIII
1	H=1							
2	Li=7	Be=9.4	B=11	C=12	N=14	O=16	F=19	
3	Na=23	Mg=24	Al=27.3	Si=28	P=31	S=32	Cl=35.5	
4	K=39	Ca=40	?=44	Ti=48	V=51	Cr=52	Mn=55	Fe=56, Co=59 Ni=59
5	Cu=63	Zn=65	?=68	?=72	As=75	Se=78	Br=80	
6	Rb=85	Sr=87	?Yt=88	Zr=90	Nb=94	Mo=96	?=100	Ru=104, Rh=104 Pd=106
7	Ag=108	Cd=112	In=113	Sn=118	Sb=122	Te=125	J=127	
8	Cs=133	Ba=137	?Di=138	?Ce=140				
9								
10			?Er=178	?La=180	Ta=182	W=184		Os=195, Ir=197 Pt=198
11	Au=199	Hg=200	Tl=204	Pb=207	Bi=208			
12				Th=231		U=240		

© Infobase Publishing

Dmitri Mendeleev's 1871 periodic table. The elements listed are the ones that were known at that time, arranged in order of increasing relative atomic mass. Mendeleev predicted the existence of elements with masses of 44, 68, and 72. His predictions were later shown to have been correct.

To place arsenic in its "proper" position, Mendeleev's first innovation was to leave two blank spaces in the table after zinc. He called the first element *eka-aluminum* and the second element *eka-silicon,* which he said corresponded to elements that had not yet been discovered but whose properties would resemble the properties of aluminum and silicon, respectively. Not only did Mendeleev predict the elements' existence, he also estimated what their physical and chemical properties should be in analogy to the elements near them. Shortly afterward, these two elements were discovered and their properties were found to be very close to what Mendeleev had predicted. Eka-aluminum was called *gallium* and eka-silicon was called *germanium.* These discoveries validated the predictive power of Mendeleev's arrangement of the elements and demonstrated that Mendeleev's periodic table could be a predictive tool, not just a compendium of information that people already knew.

The second innovation Mendeleev made involved the relative placement of tellurium and iodine. If the elements are listed in strict order of their atomic masses, then iodine should be placed before tellurium, since iodine is lighter. That would place iodine in a group with sulfur and selenium and tellurium in a group with chlorine and bromine, an arrangement that does not work for either iodine or tellurium. Therefore, Mendeleev rather boldly reversed the order of tellurium and iodine so that tellurium falls below selenium and iodine falls below bromine. More than 40 years later, after Mendeleev's death, the concept of atomic number was introduced, and it was recognized that elements should be listed in order of atomic number, not atomic mass. Mendeleev's ordering was thus vindicated, since tellurium's atomic number is one less than iodine's atomic number. Before he died, Mendeleev was considered for the Nobel Prize, but did not receive sufficient votes to receive the award despite the importance of his insights.

THE PERIODIC TABLE TODAY

All of the elements in the first 12 groups of the periodic table are referred to as *metals*. The first two groups of elements on the left-hand side of the table are the *alkali metals* and the *alkaline earth* metals. All of the alkali metals are extremely similar to each other in their chemical and physical properties, as, in turn, are all of the alkaline earths to each other. The 10 groups of elements in the middle of the periodic table are *transition metals*. The similarities in these groups are not as strong as those in the first two groups, but still satisfy the general trend of similar chemical and physical properties. The transition metals in the last row are not found in nature but have been synthesized artificially. The metals that follow the transition metals are called *post-transition* metals.

The so-called *rare earth* elements, which are all metals, usually are displayed in a separate block of their own located below the rest of the periodic table. The elements in the first row of rare earths are called *lanthanides* because their properties are extremely similar to the properties of lanthanum. The elements in the second row of rare earths are called *actinides* because their properties are extremely similar to the properties of actinium. The actinides following uranium are called *transuranium* elements and are not found in nature but have been produced artificially. The *transactinides* are elements 104 and higher that can be produced in laboratories in heavy ion collisions.

The far right-hand six groups of the periodic table—the remaining *main group* elements—differ from the first 12 groups in that more than one kind of element is found in them; in this part of the table we find metals, all of the *metalloids* (or *semimetals*), and all of the *nonmetals.* Not counting the artificially synthesized elements in these groups (elements having atomic numbers of 113 and above and that have not yet been named), these six groups contain seven metals, eight metalloids, and 16 nonmetals. Except for the last group—the *noble gases*—each individual group has more than just one kind of element. In fact, sometimes nonmetals, metalloids, and metals are all found in the same column, as are the cases with group IVB (C, Si, Ge, Sn, and Pb) and also with group VB (N, P, As, Sb, and Bi). Although similarities in chemical and physical properties are present within a column, the differences are often more striking than the similarities. In some cases, elements in the same column do have very similar chemistry. Triads of such elements include three of the *halogens* in group VIIB—chlorine, bromine, and iodine; and three group VIB elements—sulfur, selenium, and tellurium.

ELEMENTS ARE MADE OF ATOMS

An atom is the fundamental unit of matter. In ordinary chemical reactions, atoms cannot be created or destroyed. Atoms contain smaller *subatomic* particles: protons, neutrons, and electrons. Protons and neutrons are located in the *nucleus,* or center, of the atom and are referred to as *nucleons.* Electrons are located outside the nucleus. Protons and neutrons are comparable in mass and significantly more massive than electrons. Protons carry positive electrical charge. Electrons carry negative charge. Neutrons are electrically neutral.

The identity of an element is determined by the number of protons found in the nucleus of an atom of the element. The number of protons is called an element's atomic number, and is designated by the letter Z. For hydrogen, Z = 1, and for helium, Z = 2. The heaviest naturally occurring element is uranium, with $Z = 92$. The value of Z is 118 for the heaviest element that has been synthesized artificially.

Atoms of the same element can have varying numbers of neutrons. The number of neutrons is designated by the letter N. Atoms of the same element that have different numbers of neutrons are called *isotopes* of that element. The term *isotope* means that the atoms occupy the same place in the periodic table. The sum of an atom's protons and neutrons is called the atom's *mass number*. Mass numbers are dimensionless whole numbers designated by the letter A and should not be confused with an atom's *mass*, which is a decimal number expressed in units such as grams. Most elements on Earth have more than one isotope. The average mass number of an element's isotopes is called the element's *atomic mass* or *atomic weight*.

The standard notation for designating an atom's atomic and mass numbers is to show the atomic number as a subscript and the mass number as a superscript to the left of the letter representing the element. For example, the two naturally occurring isotopes of hydrogen are written ^1_1H and ^2_1H.

For atoms to be electrically neutral, the number of electrons must equal the number of protons. It is possible, however, for an atom to gain or lose electrons, forming *ions*. Metals tend to lose one or more electrons to form positively charged ions (called *cations*); nonmetals are more likely to gain one or more electrons to form negatively charged ions (called *anions*). Ionic charges are designated with superscripts. For example, a calcium ion is written as Ca^{2+}; a chloride ion is written as Cl^-.

THE PATTERN OF ELECTRONS IN AN ATOM

During the 19th century, when Mendeleev was developing his periodic table, the only property that was known to distinguish an atom of one element from an atom of another element was relative mass. Knowledge of atomic mass, however, did not suggest any relationship between an element's mass and its properties. It took several discoveries—among them that of the electron in 1897 by the British physicist John Joseph (J. J.) Thomson, *quanta* in 1900 by the German physicist Max Planck, the wave nature of matter in 1923 by the French physicist Louis de Broglie,

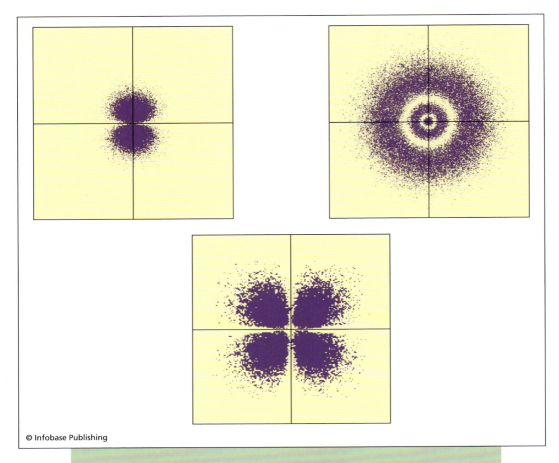

© Infobase Publishing

Hydrogen wave-function distributions for electrons in various excited states take on widely varying configurations

and the mathematical formulation of the quantum mechanical model of the atom in 1926 by the German physicists Werner Heisenberg and Erwin Schrödinger (all of whom collectively illustrate the international nature of science)—to elucidate the relationship between the structures of atoms and the properties of elements.

The number of protons in the nucleus of an atom defines the identity of that element. Since the number of electrons in a neutral atom is equal to the number of protons, an element's atomic number also reveals how many electrons are in that element's atoms. The electrons occupy regions of space that chemists and physicists call *shells.* The shells are further divided into regions of space called *subshells.* Subshells are related to angular momentum, which designates the shape of the electron orbit space around the nucleus. Shells are numbered 1, 2, 3, 4, and so forth (in theory out to infinity). In addition, shells may be designated by letters: The first shell is the "K-shell," the second shell the "L-shell," the third the "M-shell," and so forth. Subshells have letter designations, *s, p, d,* and *f* being the most common. The *n*th shell has *n* possible subshells. Therefore, the first shell has only an *s* subshell, designated *1s*; the second shell has both *s* and *p* subshells (*2s* and *2p*); the third shell *3s, 3p,* and *3d*; and the fourth shell *4s, 4p, 4d,* and *4f.* (This pattern continues for higher-numbered shells, but this is enough for now.)

An *s* subshell is spherically symmetric and can hold a maximum of two electrons. A *p* subshell is dumbbell-shaped and holds six electrons, a *d* subshell 10 electrons, and an *f* subshell 14 electrons, with increasingly complicated shapes.

As the number of electrons in an atom increases, so does the number of shells occupied by electrons. In addition, because electrons are all negatively charged and tend to repel each other *electrostatically,* as the number of the shell increases, the size of the shell increases, which means that electrons in higher-numbered shells are located, on the average, farther from the nucleus. Inner shells tend to be fully occupied with the maximum number of electrons they can hold. The electrons in the outermost shell, which is likely to be only partially occupied, will determine that atom's properties.

Physicists and chemists use *electronic configurations* to designate which subshells in an atom are occupied by electrons as well as how many electrons are in each subshell. For example, nitrogen is element number 7, so it has seven electrons. Nitrogen's electronic configuration is $1s^2 2s^2 2p^3$; a superscript designates the number of electrons that occupy a subshell. The first shell is fully occupied with its maximum of two electrons. The second shell can hold a maximum of eight electrons, but it is only partially occupied with just five electrons—two in the *2s*

ELECTRONIC CONFIGURATIONS FOR NITROGEN AND TIN

ELECTRONIC CONFIGURATION OF NITROGEN (7 ELECTRONS)

Energy Level	Shell	Subshell	Number of Electrons
1	K	s	2
2	L	s	2
		p	3
			7

ELECTRONIC CONFIGURATION OF TIN (50 ELECTRONS)

Energy Level	Shell	Subshell	Number of Electrons
1	K	s	2
2	L	s	2
		p	6
3	M	s	2
		p	6
		d	10
4	N	s	2
		p	6
		d	10
5	O	s	2
		p	2
			50

subshell and three in the *2p*. Those five outer electrons determine nitrogen's properties. For a heavy element like tin (Sn), electronic configurations can be quite complex. Tin's configuration is $1s^2 2s^2 2p^6 3s23p^6 4s^2 3d^{10} 4p^6 5s^2 4d^{10} 5p^2$ but is more commonly written in the shorthand notation [Kr] $5s^2 4d^{10} 5p^2$ where [Kr] represents the electron configuration pattern for the noble gas Krypton. (The pattern continues in this way for shells with higher numbers.) The important thing to notice about tin's configuration is that all of the shells except the last one are fully occupied. The fifth shell can hold 32 electrons, but in tin there are only four electrons in the fifth shell. The outer electrons determine an element's properties. The table on the previous page illustrates the electronic configurations for nitrogen and tin.

ATOMS ARE HELD TOGETHER WITH CHEMICAL BONDS

Fundamentally, a *chemical bond* involves either the sharing of two electrons or the transfer of one or more electrons to form ions. Two atoms of nonmetals tend to share pairs of electrons in what is called a *covalent* bond. By sharing electrons, the atoms remain more or less electrically neutral. However, when an atom of a metal approaches an atom of a nonmetal, the more likely event is the transfer of one or more electrons from the metal atom to the nonmetal atom. The metal atom becomes a positively charged ion and the nonmetal atom becomes a negatively charged ion. The attraction between opposite charges provides the force that holds the atoms together in what is called an *ionic* bond. Many chemical bonds are also intermediate in nature between covalent and ionic bonds and have characteristics of both types of bonds.

IN CHEMICAL REACTIONS, ATOMS REARRANGE TO FORM NEW COMPOUNDS

When a substance undergoes a *physical change,* the substance's name does not change. What may change is its temperature, its length, its *physical state* (whether it is a solid, liquid, or gas), or some other characteristic, but it is still the same substance. On the other hand, when a substance undergoes a *chemical change,* its name changes; it is a different substance. For example, water can decompose into hydrogen gas and oxygen gas, each of which has substantially different properties from water, even though water is composed of hydrogen and oxygen atoms.

In chemical reactions, the atoms themselves are not changed. Elements (like hydrogen and oxygen) may combine to form compounds (like water), or compounds can be decomposed into their elements. The atoms in compounds can be rearranged to form new compounds whose names and properties are different from the original compounds. Chemical reactions are indicated by writing chemical equations such as the equation showing the decomposition of water into hydrogen and oxygen: $2 H_2O$ (l) $\rightarrow 2 H_2$ (g) $+ O_2$ (g). The arrow indicates the direction in which the reaction proceeds. The reaction begins with the *reactants* on the left and ends with the *products* on the right. We sometimes designate the physical state of a reactant or product in parentheses—"s" for solid, "l" for liquid, "g" for gas, and "aq" for *aqueous* solution (in other words, a solution in which water is the *solvent*).

IN NUCLEAR REACTIONS THE NUCLEI OF ATOMS CHANGE

In ordinary chemical reactions, chemical bonds in the reactant species are broken, the atoms rearrange, and new chemical bonds are formed in the product species. These changes only affect an atom's electrons; there is no change to the nucleus. Hence there is no change in an element's identity. On the other hand, nuclear reactions refer to changes in an atom's nucleus (whether or not there are electrons attached). In most nuclear reactions, the number of protons in the nucleus changes, which means that elements are changed, or *transmuted,* into different elements. There are several ways in which transmutation can occur. Some transmutations occur naturally, while others only occur artificially in nuclear reactors or particle accelerators.

The most familiar form of transmutation is *radioactive decay,* a natural process in which a nucleus emits a small particle or *photon* of light. Three common modes of decay are labeled *alpha, beta,* and *gamma* (the first three letters of the Greek alphabet). Alpha decay occurs among elements at the heavy end of the periodic table, basically elements heavier than lead. An alpha particle is a nucleus of helium 4 and is symbolized as $_2^4\text{He}$ or α. An example of alpha decay occurs when uranium 238 emits an alpha particle and is changed into thorium 234 as in the following reaction: $_{92}^{238}\text{U} \rightarrow {_2^4}\text{He} + {_{90}^{234}}\text{Th}$. Notice that the *parent* isotope, U-238, has 92 protons, while the *daughter* isotope, Th-234, has only 90 protons. The decrease in the number of protons means a change in the identity of the element. The mass number also decreases.

Any element in the periodic table can undergo beta decay. A beta particle is an electron, commonly symbolized as β⁻ or e⁻. An example of beta decay is the conversion of cobalt 60 into nickel 60 by the following reaction: $_{27}^{60}\text{Co} \rightarrow {_{28}^{60}}\text{Ni} + e^-$. The atomic number of the daughter isotope is one greater than that of the parent isotope, which maintains charge balance. The mass number, however, does not change.

In gamma decay, photons of light (symbolized by γ) are emitted. Gamma radiation is a high-*energy* form of light. Light carries neither mass nor charge, so the isotope undergoing decay does not change identity; it only changes its energy state.

Elements also are transmuted into other elements by nuclear *fission* and *fusion*. Fission is the breakup of very large nuclei (at least as heavy as uranium) into smaller nuclei, as in the fission of U-236 in the following reaction: $_{92}^{236}\text{U} \rightarrow {_{36}^{94}}\text{Kr} + {_{56}^{139}}\text{Ba} + 3n$, where n is the symbol for a neutron (charge = 0, mass number = +1). In fusion, nuclei combine to form larger nuclei, as in the fusion of hydrogen isotopes to make helium. Energy may also be released during both fission and fusion. These events may occur naturally—fusion is the process that powers the Sun and all other stars—or they may be made to occur artificially.

Elements can be transmuted artificially by bombarding heavy target nuclei with lighter projectile nuclei in reactors or accelerators. The transuranium elements have been produced that way. Curium, for example, can be made by bombarding plutonium with alpha particles. Because the projectile and target nuclei both carry positive charges, projectiles must be accelerated to velocities close to the speed of light to overcome the force of repulsion between them. The production of successively heavier nuclei requires more and more energy. Usually, only a few atoms at a time are produced.

ELEMENTS OCCUR WITH DIFFERENT RELATIVE ABUNDANCES

Hydrogen overwhelmingly is the most abundant element in the universe. Stars are composed mostly of hydrogen, followed by helium and only very small amounts of any other element. Relative abundances of elements can be expressed in parts per million, either by mass or by numbers of atoms.

On Earth, elements may be found in the *lithosphere* (the rocky, solid part of Earth), the *hydrosphere* (the aqueous, or watery, part of Earth), or the atmosphere. Elements such as the noble gases, the rare earths, and commercially valuable metals like silver and gold occur in only trace quantities. Others, like oxygen, silicon, aluminum, iron, calcium, sodium, hydrogen, sulfur, and carbon are abundant.

HOW NATURALLY OCCURRING ELEMENTS HAVE BEEN DISCOVERED

For the elements that occur on Earth, methods of discovery have been varied. Some elements—like copper, silver, gold, tin, and lead—have been known and used since ancient or even prehistoric times. The origins of their early *metallurgy* are unknown. Some elements, like phosphorus, were discovered during the Middle Ages by alchemists who recognized that some mineral had an unknown composition. Sometimes, as in the case of oxygen, the discovery was by accident. In other instances—as in the discoveries of the alkali metals, alkaline earths, and lanthanides—chemists had a fairly good idea of what they were looking for and were able to isolate and identify the elements quite deliberately.

To establish that a new element has been discovered, a sample of the element must be isolated in pure form and subjected to various chemical and physical tests. If the tests indicate properties unknown in any other element, it is a reasonable conclusion that a new element has been discovered. Sometimes there are hazards associated with isolating a substance whose properties are unknown. The new element could be toxic, or so reactive that it can explode, or be extremely radioactive.

During the course of history, attempts to isolate new elements or compounds have resulted in more than just a few deaths.

HOW NEW ELEMENTS ARE MADE

Some elements do not occur naturally, but can be synthesized. They can be produced in nuclear reactors, from collisions in particle accelerators, or can be part of the *fallout* from nuclear explosions. One of the elements most commonly made in nuclear reactors is technetium. Relatively large quantities are made every day for applications in nuclear medicine. Sometimes, the initial product made in an accelerator is a heavy element whose atoms have very short *half-lives* and undergo radioactive decay. When the atoms decay, atoms of elements lighter than the parent atoms are produced. By identifying the daughter atoms, scientists can work backward and correctly identify the parent atoms from which they came.

The major difficulty with synthesizing heavy elements is the number of protons in their nuclei (Z > 92). The large amount of positive charge makes the nuclei unstable so that they tend to disintegrate either by radioactive decay or *spontaneous fission*. Therefore, with the exception of a few transuranium elements like plutonium (Pu) and americium (Am), most artificial elements are made only a few atoms at a time and so far have no practical or commercial uses.

THE NONMETALS CORNER OF THE PERIODIC TABLE

The designated nonmetals in this volume are as follows:

- Hydrogen
- Carbon
- Nitrogen
- Oxygen
- Phosphorus
- Sulfur
- Selenium

The following is the key to understanding each element's information box that appears at the beginning of each chapter.

Information box key. E represents the element's letter notation (for example, H = hydrogen), with the Z subscript indicating proton number. Orbital shell notations appear in the column on the left. For elements that are not naturally abundant, the mass number of the longest-lived isotope is given in brackets. The abundances (atomic %) are based on meteorite and solar wind data. The melting point (M.P.), boiling point (B.P.), and critical point (C.P.) temperatures are expressed in degrees Celsius. Should sublimation and critical point temperatures apply, these are indicated by s and t, respectively.	**Element**
	K $\qquad$ M.P.°
	L $\quad$ **E**$_z$ $\quad$ B.P.°
	M $\qquad$ C.P.°
	N
	O
	P $\quad$ Oxidation states
	Q $\quad$ Atomic weight
	$\quad$ Abundance%

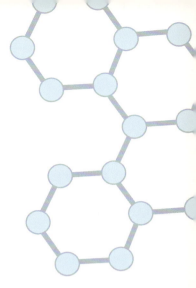

1

Hydrogen: Ubiquitous by Nature

Hydrogen—element number 1—is the lightest and most abundant element in the universe. In fact, 93 percent of all the atoms in the universe are hydrogen atoms. Hydrogen is the primary fuel of the stars, a major component of water, and an important element in the molecules constituting the bodies of living organisms. The story of hydrogen begins with the *big bang*. It continues with the identification of hydrogen as an element, on to the first hydrogen-filled balloon, and into the present day with the interest in hydrogen as a clean, renewable source of energy. This chapter explores the origin of hydrogen atoms (which ultimately led to the origin of all matter in the universe), the discovery of hydrogen on Earth, current research, and many of hydrogen's modern-day uses.

THE BASICS OF HYDROGEN

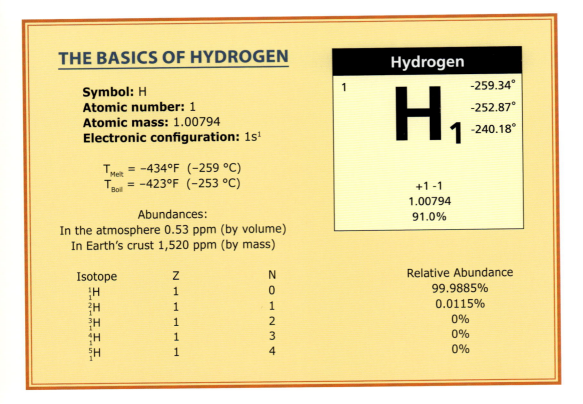

Symbol: H
Atomic number: 1
Atomic mass: 1.00794
Electronic configuration: $1s^1$

$T_{Melt} = -434°F$ $(-259 °C)$
$T_{Boil} = -423°F$ $(-253 °C)$

Abundances:
In the atmosphere 0.53 ppm (by volume)
In Earth's crust 1,520 ppm (by mass)

Isotope	Z	N	Relative Abundance
^1_1H	1	0	99.9885%
^2_1H	1	1	0.0115%
^3_1H	1	2	0%
^4_1H	1	3	0%
^5_1H	1	4	0%

Within the element box:

Hydrogen
1   H   1
-259.34°
-252.87°
-240.18°
+1 -1
1.00794
91.0%

THE ASTROPHYSICS OF HYDROGEN

Although hydrogen is the most abundant element in the universe, it was not created spontaneously during the explosion that began our universe 15 billion years ago. The big bang formed a chaotic mixture of *matter, antimatter,* and *radiation.* Antimatter meeting matter underwent mutually explosive annihilation, becoming energy that could be absorbed by the subatomic particles created in the blast. If the amount of antimatter had equaled the amount of matter, everything would have been annihilated within a tenth of a second. Fortunately, matter was a slightly larger portion of the stew, and the entire system cooled sufficiently that some of the matter could form *nucleons*—the collective name given to the neutrons and protons that form the cores of atoms. Several hundred thousand years had to pass before free-flying electrons, attracted to the positively charged protons, could remain attached and atoms were born. Elegant in its simplicity, hydrogen was the most easily formed and remains the dominant atomic species in the universe today.

But where is it? Hydrogen composes a minuscule portion of our atmosphere—only one part per million. Although it is bound in the molecules of our oceans and rivers, hydrogen does not exist in its pure molecular form in very many places on Earth.

To find where the majority of hydrogen is located, scientists have examined *spectroscopic* data from stars. In the early universe, currents and spirals formed from matter attracted to other matter via the gravitational force, initiating clouds of hydrogen atoms. This activity still goes on today. Once a cloud reaches a temperature around 5 million K

Hydrogen nuclei fuse to make helium nuclei in the core of the Sun, producing energy in the form of electromagnetic radiation. (*Extreme Ultraviolet Imaging Telescope Consortium/NASA*)

and a density about 100 times that of water, the hydrogen *nuclei* begin to fuse into nuclei of helium. For each *fusion* event, about 5×10^{-12} *joules* of energy are released—a very small amount, but the huge number of fusion reactions occurring each second results in the Sun radiating energy at the astonishing rate of 4×10^{26} *watts*. It is this process of hydrogen fusing into helium in the Sun that is the source of light, heat, and ultimately all the fundamentally useable energy available on Earth.

The only way scientists know that the Sun is mostly hydrogen is from experiments performed here on Earth. Each element in the periodic table has a distinct signature, called its *spectrum*. Electrons do not stick to the nuclei in atoms, but surround the core in a fashion that scientists have modeled variously as orbits, clouds, and probability densities. More detail will unfold later in this chapter, but for the moment, the model of an atom to be pictured is that of an electron orbiting the nucleus like the Moon orbits Earth.

An electron, however, unlike the Moon, can have many different orbits if it absorbs the right amount of energy. That energy can come from collisions with other particles. The atom can also absorb light, allowing the electron to jump to a higher-energy orbit, but it does not tend to remain in that excited state. It will eventually relax back to its least energetic state. When the electron drops down from a higher to a lower energy state, it gives off energy in the form of light or, more precisely, *electromagnetic radiation*. The human eye cannot see all *wavelengths* of electromagnetic radiation, but there are instruments called *spectroscopes* that can detect and measure the radiation. Hydrogen electrons emit different wavelengths than do helium electrons, which emit different wavelengths than, say, carbon. In fact, every element has a unique *spectrum* by which one can identify it.

This can be observed in the laboratory (even a high school physics laboratory) using easily obtainable tubes of atomic gases. Scientists look at the spectra from the heated tubes of gas in the lab and then compare these to spectra from stars. Most of our stars are moving away from us as the universe expands, so we need to include a *redshift factor,* but the patterns remain the same.

All the hydrogen there is—and all other naturally occurring elements—are produced in stars. Most Earth-based elements heavier than iron, however, must have been created in stellar *supernova* events, a phenomenon to be discussed in chapter 4.

DISCOVERY AND NAMING OF HYDROGEN

Hydrogen is the most abundant element in the universe and the tenth most abundant element in Earth's crust. Hydrogen atoms make up 93 percent of all atoms in the universe. About 6.5 percent of the atoms in the universe are helium atoms. The remaining scant 0.5 percent of the universe consists of the atoms of all of the other elements, and yet some of those elements were isolated and identified by ancient people. Why then were other, less abundant, substances recognized as elements so much earlier than hydrogen was? The simplest answer is probably that hydrogen is a colorless, odorless gas. In ancient times, Greek philosophers thought there were only four elements—earth, air, fire, and water—and that all other substances were mixtures of those four elements. Scientists no longer classify any of these as elements. Later, even the elements that were known to medieval alchemists were a liquid (mercury) and solids (such as gold). Since hydrogen is a colorless and odorless gas, alchemists did not know to look for it, so it seems natural that the discovery of hydrogen—and gases like oxygen and nitrogen—came after the Middle Ages.

Suspicion of hydrogen's existence dates to 1671, when the English natural philosopher Robert Boyle (1627–91) noted the flammability of the gas that results from the reaction of iron with hydrochloric acid. Boyle, however, did not identify the fumes he obtained as being those of a new element. Credit for the discovery of hydrogen goes to the English chemist Henry Cavendish (1731–1810). There were alchemists before Cavendish who had dissolved metals in acids and observed hydrogen and noted its flammability, but Cavendish, in 1766, was the first person to state that hydrogen was different from all other gases. He called the new gas "inflammable air from the metals." He was also the first person to obtain pure samples of hydrogen and to describe its low density. Cavendish dissolved metals such as zinc, iron, and tin in hydrochloric and sulfuric acid to isolate hydrogen. He made the mistake, however, of thinking that hydrogen was a component of the metals, whereas shortly afterward scientists recognized that hydrogen gas came not from the metals, but from the acids used to dissolve the metals.

Following the discovery of oxygen in 1774, Cavendish combined hydrogen and oxygen to make water. This experiment showed that water is a compound, not an element, sounding the death knell to the Greek notion of the four elements earth, air, fire, and water. It was the French chemist, Antoine-Laurent Lavoisier (1743–94), who gave the name *hydrogen* to Cavendish's "inflammable air" after hearing of Cavendish's success at making water from hydrogen and oxygen. The word *hydrogen* means "water producer."

A PLANETARY NOTION: THE BOHR MODEL

Since antiquity, natural philosophers have debated questions regarding the fundamental nature of matter. Is matter infinitely divisible, or does there exist a smallest possible indivisible piece of matter? Democritus, the ancient Greek philosopher, taught that matter consists of fundamental particles that cannot be further cut or subdivided. The word "atom" today stems from Democritus's teaching: *a* = not, *tom* = cut. Aristotle (384–322 B.C.E.), however, believed that matter is infinitely divisible; there are no fundamental particles that cannot be further subdivided. Since Aristotle was the most influential natural philosopher of ancient times, atomic theory was largely abandoned until the beginning of the 19th century.

In 1808, the English chemist John Dalton (1766–1844) published *A New System of Chemical Philosophy,* in which he revived the atomic theory of Democritus and showed that then-current knowledge in chemistry was consistent with the theory that matter is composed of atoms. Although atoms are too small to be seen by the naked eye—or even by optical microscopes—atomic theory came to be accepted during the 19th century as the best explanation for the fundamental composition of matter. In recognition of his contributions, Dalton is often referred to as the "father of modern atomic theory."

Throughout the 19th century, however, little progress was made regarding the possible composition or structure of atoms. Imagined as

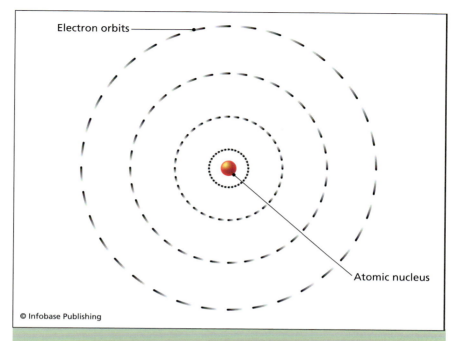

Danish physicist Niels Bohr postulated that the electron in hydrogen travels around the nucleus in the manner in which a planet orbits the Sun. Hence his model was called the "planetary model."

little hard spheres, the atoms of different elements were known to differ only in their relative masses. This view changed with the discovery of radioactivity in 1896 by the French physicist, Henri Becquerel (1852–1908), and the discovery of the electron in 1897 by the English physicist, J. J. Thomson (1856–1940). In addition, physicists soon began talking about atoms containing fundamental positively charged particles. After the discovery of atomic nuclei, the nucleus of the lightest isotope of hydrogen was called a proton. (Although scientists postulated the existence of electrically neutral particles in the early 1920s, the discovery of the neutron eluded physicists until 1932.) These discoveries suggested that atoms were not hard spheres; however, a more detailed model of the atom was needed. Being English, Thomson was well acquainted with the traditional yuletide dessert plum pudding, which consists mostly of raisins, plums, and spices. By way of analogy Thomson suggested a "plum-pudding model" of the atom in which the electrons resembled the relatively small raisins in the pudding, and the positively charged particles resembled the much larger plums. Thomson's model was a beginning in the attempt to visualize the internal structure of atoms. Such a homogeneous model of the atom, however, offered little insight into any relationship between the structures of atoms and their properties and was later replaced with more detailed models.

In 1911, the British physicist and Nobel laureate Ernest Rutherford (1871–1937) published the article "The Scattering of Alpha and Beta Particles by Matter and the Structure of the Atom" in *Philosophical Magazine.* In this article, Rutherford reported the results of an experiment that demonstrated that the protons and electrons in atoms are not distributed homogeneously. Instead, the protons are concentrated in a relatively tiny region Rutherford called the nucleus (from the Latin, meaning "kernel"). The electrons are *extranuclear;* electrons are located in a relatively much larger volume of space surrounding the nucleus. Rutherford's discovery of the nucleus was immediately accepted within the scientific community. However, the relationship, if any, between atomic structure and properties was still unclear.

For several decades prior to 1910, scientists had been studying the spectra of elements and compounds. When white light passes through a sample of a cool gas or liquid, certain frequencies are absorbed, leading to an *absorption spectrum* consisting of dark lines embedded in the rainbow colors of the spectrum. When a gas is heated, the gas emits certain *frequencies* of light, leading to an *emission spectrum* consisting of bright lines against a dark background. (Similar results can be obtained using any frequency of electromagnetic radiation. Spectra outside the visible region must be measured by instruments sensitive to those particular regions of the spectrum.)

Scientists recognized that an element's spectrum is like human fingerprints; just as each person has unique fingerprints, each element in the periodic table has a unique spectrum. The spectrum of a compound is simply a combination of the spectra of the elements in that compound. The uniqueness of spectra makes them powerful analytical tools in the identification of the elements found in minerals, in biological samples, in Earth's atmosphere, and in the atmospheres of stars. Although much *empirical* data had been accumulated, there was no theoretical framework that explained or predicted spectra.

Finally, an explanation was provided in 1913 by Niels Bohr (1885–1962), a Danish physicist who tackled the problem of trying to understand fundamental atomic structure. Bohr postulated that the electron in hydrogen travels around the nucleus in the manner in which a planet orbits the Sun. Hence his model was called the "planetary model." The important distinction between the orbit of a planet around the Sun and the orbit of an electron around a nucleus is that the distance of a planet from the Sun is arbitrary, whereas in the Bohr model an electron cannot exist at just any distance from a nucleus. An electron can orbit the nucleus only at particular fixed, or discrete, distances from the nucleus.

The energy of the electron in orbit around the nucleus is the sum of the electron's *kinetic* and *electrostatic potential* energies. Since the potential energy depends on the distance the electron is from the nucleus, and in Bohr's model distances can only be discrete quantities, the model dictates that the total energy of the electron can only take on discrete values. An electron seems to jump from one orbit to another, but it cannot exist in a stable state in the region between two orbits. Orbits close to the nucleus represent states of lower energy. As the distance of an orbit from the nucleus increases, the energy of the orbit increases. When an electron occupies the orbit closest to the nucleus, the electron is said to be in its ground state, or state of lowest energy. When an electron occupies an orbit farther from the nucleus, the electron is said to be in an excited state, or state of higher energy.

An electron can make the transition from a state of lower energy to a state of higher energy by absorbing a photon of light. However, it cannot absorb a photon possessing just any frequency (which, in the visible region, determines the color of the light). The energy of the photon is proportional to its frequency. This energy must correspond precisely to the difference between the electron's higher and lower energy states. Because there are only a finite number of possible energy states, only photons with specific energies can be absorbed. Therefore, only very specific absorption lines are observed in an element's or compound's spectrum. (In the reverse process, when an

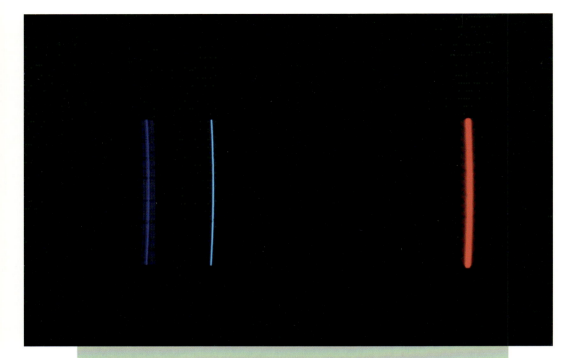

Emission spectrum of hydrogen. When atoms of an element are excited (e.g., by heating), they return to their state of lowest energy by emitting radiation at specific wavelengths. If this radiation is passed through a spectrometer, a spectrum is produced that displays the element's characteristic emission lines. The lines are a unique "fingerprint" of an element. (*Department of Physics, Imperial College London/Photo Researchers, Inc.*)

electron makes the transition from higher to lower energy, a photon of light must be emitted. Various transitions provide a specific set of emission lines in an element's or compound's spectrum.)

The planetary model of the hydrogen atom seemed at first to successfully explain the spectrum that is observed in the *ultraviolet,* visible, and *infrared* radiation regions. Qualitatively, Bohr's planetary model provides a reasonable explanation for the origin of spectral lines for all elements. Quantitatively, however, Bohr's model provides only approximate results for the hydrogen atom, but wildly incorrect results for all other atoms. These discrepancies were soon discovered. One, involving more precise measurements of the frequencies of hydrogen's spectral lines, showed that the experimental frequencies did not exactly match the frequencies predicted by Bohr's model. Another discrepancy was that Bohr's model did not predict the effects of electric or magnetic fields on hydrogen's spectrum. Nevertheless, because Bohr's model incorporated *quantum theory* into its formulation, it was an extremely important step in the ultimate elucidation of atomic structure and it garnered Bohr the Nobel Prize in 1922. It was a few years later that a more detailed treatment of the atom using quantum theory was presented.

A QUANTUM SOLUTION

Although Bohr's model was successful at predicting spectral lines, it only applied to one-electron atoms and provided no way to calculate

Niels Bohr was a Danish physicist who tackled the problem of trying to understand fundamental atomic structure. *(AP Photo/Alan Richard)*

transition rates. Transition rates were known to vary, as some spectral lines were considerably brighter than others, indicating that some energy jumps occurred much more frequently. The model also allowed only for circular paths, an unstable condition for an orbiting electron. In addition, Bohr's model required that the classical orbital angular moment (a product of the linear momentum vector and the radius vector) be equal in magnitude to an integer multiple of *Planck's constant* divided by the quantity 2π. This notion seemed to defy common sense, as it was unnatural in classical mechanics to have physical quantities that had to be whole-number multiples of another quantity.

Louis de Broglie, then a doctoral student at Paris University, knew that whole-number behavior in physics was commonly associated with periodicity in a system. Perhaps a periodic nature had to be a property of electrons in atoms. Since waves are the quintessential example of periodicity, he hypothesized that not only did light have a wave nature, but so must electrons. De Broglie pictured electrons as particles embedded in *standing waves* around a nucleus. If the two modes of existence were to mesh, he had somehow to relate wavelength to mass or momentum. His famous formula $\lambda = h/p$ (where λ is the wavelength, p is the particle momentum, and h is *Planck's constant*) had the right dimensions, but needed to be confirmed by experimental tests.

To find out if electrons present a diffraction pattern as light waves do, they had to be beamed through an opening of 10^{-10}m—about 1/10,000 the radius of a human hair. It is impossible to machine such a narrow slit, but in 1927, physicists Clinton Davisson and Lester Germer at the Bell Laboratories in New Jersey were able to shoot electrons through a nickel crystal, whose structure is arranged by nature to have atomic planes just the right distance apart. They heated a metal filament to eject electrons that then traveled through the crystal to land in an electron detector. If electrons did not have a wave nature, the distribution of electron hits would have been uniform, but it was not. Instead, a distinct pattern of light and dark, like the crests and troughs of a wave, was observed, confirming de Broglie's controversial idea.

De Broglie's formula was useful for knowing a particle's wavelength, but a more comprehensive understanding of the electron's motion and location while bound to a nucleus was needed. In order to be able to

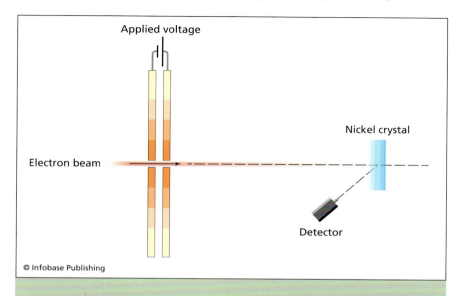

To find out if electrons present a diffraction pattern as light waves do, physicists Clinton Davisson and Lester Germer at the Bell Laboratories in New Jersey in 1927 were able to "shoot" electrons through a nickel crystal, whose structure is arranged by nature to have atomic planes just the right distance apart. They heated a metal filament to eject electrons that then traveled through the crystal to land in an electron detector.

predict behavior such as the flux of electrons (rate of flow per unit area) in a scattering experiment like that of Davisson and Germer, one needs a mathematical function to describe electron motion. Austrian physicist Erwin Schrödinger took on the challenge by adapting the known wave equation for light (electromagnetic waves) to make it comply with the law of energy conservation, free particle motion, and de Broglie's calculation for wavelength. (Eventually Schrödinger had to modify the equation to accommodate a *spin* nature to the electron as tested by observing atoms in a magnetic field.)

The solution to Schrödinger's equation is a *wave function*. The solution provides a way to calculate the probability of finding the electron in a particular region near the nucleus for a given energy,

(continued on page 16)

HEAVY HYDROGEN: DEUTERIUM, TRITIUM, AND BEYOND

Although hydrogen is known as the lightest element, it has isotopes that defy this description. From deuterium to ^5H, hydrogen with added neutrons has fascinating properties that assist research and technology in many regimes.

Researchers have been interested in heavy hydrogen since the early 1900s when Harold Urey, a chemistry professor at Columbia University, developed a method of distilling liquid hydrogen to make deuterium—hydrogen with a neutron connected to the proton—an achievement for which he won the 1934 Nobel Prize in chemistry.

When used in place of hydrogen, deuterium or ^2H (sometimes designated as D) results in water approximately 10 percent denser than normal. Termed "heavy water," D_2O is harmless in small doses and can therefore be used safely as a tracer in the body, most commonly in measuring a subject's *metabolic rate*. Heavy water is also used as a neutron *moderator*, meaning it is able to slow neutrons by collisions without absorbing them. This process is crucial for the chain reaction in nuclear reactors, where fast neutrons are produced by the fission process, but slow or *thermal* neutrons are more likely to induce fission.

The universal abundance of deuterium is the subject of ongoing investigation. All deuterium nuclei formed a few minutes after the big bang, providing the basis for the heavier elements. It is understood to be consumed solely by stellar burning—the nuclear fusion process in stars. New observational results, however, show that universal abundance is about 20 percent more than projected under this scenario. This offers spectroscopists and theorists new channels of exploration.

Hydrogen with two neutrons, named tritium (^3H or T), is less stable than the lighter isotopes and is radioactive, with a half-life of 12.26 years. Produced by cosmic ray protons colliding with nitrogen in the upper atmosphere, trace amounts are found in air and less abundantly in water. Tritium can also be made in the laboratory by bombarding lithium atoms with neutrons, which is the source of tritium for use in fusion energy research.

Deuterium-tritium fusion is considered the most likely process to result in a fusion reactor suitable for electricity production. In this process, a confined gas of deuterium and tritium atoms must be heated to nearly 100,000,000 K. Each fusion of D with T produces a helium nucleus, or *alpha particle,* and a neutron. The 17.6 million electron volts (MeV) of energy released per reaction are substantial, but only 3.5 MeV are carried away by the charged particle—the more useable form of energy for electricity. Although the method is well understood, it is still highly inefficient; much more energy must be put into the process than is produced. Fusion is not expected to be a viable source of power for humankind for at least the next 50 years.

The heavier isotopes of hydrogen, ^4H and ^5H, have been produced in the laboratory via tritium collisions with deuterium and tritium, respectively. Extremely short-lived (on the order of 10^{-23} seconds), these atoms most likely also exist in extreme temperature and density regimes in stars, though they cannot be detected spectroscopically.

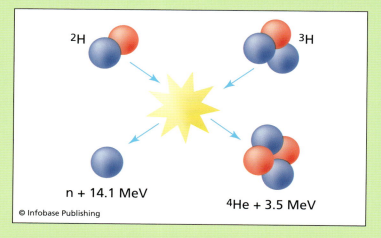

© Infobase Publishing

Schematic of deuterium-tritium fusion

(continued from page 13)

angular momentum, and spin state as well as the probability for a jump between two states. It is important to note that an exact solution for Schrödinger's equation is known only for one-electron atoms where no additional forces exist from further electrons. All solutions for multi-electron atoms require extensive approximation. For that reason, experimentation leads theory in the field of atomic physics.

Conceptualizing this quantum mechanical behavior can be a problem. Schrödinger's equation implies that the electron cannot be found at a specific location; rather, there is some probability (less than 100 percent) that it may be found there. This uncertain idea of trying to figure out the odds of locating a particle led Einstein to state in a 1926 letter to Max Born, "I am convinced that He (God) does not play dice." Wave mechanics, however, is the best description yet found for observed electron distributions in atoms.

THE NEGATIVE HYDROGEN ION

The negative hydrogen ion (H^-), or hydrogen with an extra electron attached, is a glaring example of the necessity of quantum mechanical theory to describe atomic behavior. Classical physics, considering only electrostatic forces among the three charged particles, predicts that this ion should not exist in a stable, bound state. Yet H^- has been observed experimentally for decades.

It is only possible to understand how two electrons can be bound to one proton by considering the electron wave functions. In quantum mechanics, the electrons cannot be modeled as pointlike particles orbiting the nucleus, but must be pictured as fuzzy distributions of probability. In H^-, the electrons are in close enough proximity that their probability distributions, or wave functions, overlap. This overlap induces a positive correlation that allows the bound state of the ion. This means that the electrons do not have simple individual independent wave functions, but share a different and more complicated wave function.

Electron correlation is a relatively new concept and has been studied by monitoring the *doubly excited states* of this simplest atomic three-body problem. Unlike the hydrogen atom, H^- has no singly excited state where just one electron jumps to a higher level. Instead, when sufficient energy is introduced into the negative ion (via collision with other particles or excitation by light), both electrons are simultaneously excited to higher levels. These so-called doubly excited states are evidence of a rather alarming phenomenon—it has been demonstrated that the electrons can actually share the energy of a single photon. This took physicists by surprise because a photon was understood to be an indivisible packet of energy and should be absorbed as such. Work continues on doubly and even triply excited states of various atoms, mainly using electron storage rings—particle accelerators that produce *synchrotron radiation* to facilitate excitation of the electrons.

THE CHEMISTRY OF HYDROGEN

Hydrogen itself is easily obtainable from the *electrolysis* of water. In an electrolytic cell, two terminals—an *anode* and a *cathode*—are connected to a power supply. When an electrical current is sent through the cell, bubbles of hydrogen gas are formed at the cathode and bubbles of oxygen gas are formed at the anode. The net chemical equation for the reaction is

$$2\ H_2O\ (l) \rightarrow 2\ H_2\ (g) + O_2\ (g).$$

Note that in water the hydrogen atoms are bonded covalently to an oxygen atom. In molecular hydrogen, the hydrogen atoms are bonded covalently to each other. Hydrogen gas itself is highly flammable; in the reverse of the reaction above, H_2 and O_2 recombine to form water, releasing energy, sometimes explosively.

Chemical bonding involves an atom's electrons. Electrons can be transferred from one atom to another (making an ionic bond), or electrons can be shared between two atoms (making a covalent bond). When an atom loses one or more electrons, it is left with a net positive charge, and the resulting species is called a positive ion, or cation. When an atom gains one or more electrons, it assumes a net negative charge, and the resulting species is called a negative ion, or anion. When an atom more or less equally shares electrons with another atom, no ion is formed. As a general rule, the chemistry of metallic elements is dominated by the tendency of the atoms of metals to form almost exclusively cations, whereas the chemistries of metalloids and nonmetals are dominated by the tendency of their atoms to form either anions or covalent bonds, but rarely positive ions.

Hydrogen is more versatile than most elements in that hydrogen can form all three: cations, anions, and covalent bonds. The tendency of hydrogen to form positive ions, as metals do, is responsible for its usual placement in the periodic table at the top of the first column above lithium. Some periodic tables, however, show hydrogen at the top of the column of halogens, reflecting hydrogen's ability also to form negative ions. Some periodic tables even place hydrogen in both positions.

There are various definitions of acids and bases. The one used here is attributed to a theory developed in 1923 independently by Johannes Brønsted (1879–1947), a Danish chemist, and Thomas Lowry (1874–1936), a British chemist. Recall that an atom of ordinary hydrogen has only a proton and an electron, and no neutrons. Therefore, a cation of ordinary hydrogen (H^+) is just a proton. In the Brønsted-Lowry definition of acids and bases, an acid is a proton donor, that is, it can react with other compounds or ions by transferring one or more H^+ ions to the other compounds or ions. A base is a proton acceptor: It can react with the H^+ ions of compounds or ions that are acids. Some chemical species, such as H_2O, are said to be *amphiprotic,* that is, they are both donors and acceptors of protons; they are both an acid and a base. These definitions are illustrated in the following examples ("aq" means the reaction is taking place in aqueous solution):

$$HCl \text{ (aq)} + H_2O \text{ (l)} \rightarrow Cl^- \text{ (aq)} + H_3O^+ \text{ (aq)}.$$

$$(\text{acid} + \text{base} \rightarrow \text{base} + \text{acid})$$

$$NH_3 \text{ (aq)} + H_2O \text{ (l)} \rightarrow NH_4^+ \text{ (aq)} + OH^- \text{ (aq)}.$$

$$(\text{base} + \text{acid} \rightarrow \text{acid} + \text{base})$$

In the first equation, hydrochloric acid (HCl) is a proton donor and water is a proton acceptor. Therefore, in that equation, HCl is an acid and water is a base. In the second equation, ammonia (NH_3) is a proton acceptor and water is a proton donor. Therefore, in that equation, NH_3 is a base and water is an acid. The ability of water to behave as an acid in one reaction or as a base in a different reaction illustrates water's amphiprotic nature. Notice that in an acid-base reaction, the reactant that is an acid is changed into a product that is a base; the reactant that is a base is changed into a product that is an acid. These combinations are called acid-base *conjugate pairs.* Thus, HCl and Cl^- form an acid-base conjugate pair, and NH_4^+ and NH_3 form a conjugate pair.

A slightly different but equivalent definition of acids and bases is suggested by these equations. The first reaction results in an increase in hydronium ions (H_3O^+). The second reaction results in an increase in hydroxide ions (OH^-). Therefore, an acid can be defined as a chemical substance that, when added to water, results in an increase in the concentration of hydronium ions. A base is a chemical substance that, when added to water, results in an increase in the concentration of hydroxide ions. Hydronium ions give acid solutions the properties we associate with acids—sour taste, corrosiveness, and the ability to turn blue *litmus* paper red. Hydroxide ions give basic (or *alkaline*) solutions the properties of feeling "soapy," causticity, and the ability to turn red litmus paper blue.

Compounds of hydrogen with other elements are typically called *hydrides.* In metal hydrides, the hydrogen is a negative ion. With metalloids and nonmetals, the hydrides can involve covalent bonding or the formation of the positive hydrogen ion. When hydrogen combines with nitrogen, ammonia (NH_3) is formed. Ammonia is a base. Hydrogen combines with phosphorus to form phosphine (PH_3). Water (H_2O) is both an acid and a base. The hydrides of the other elements in oxygen's family are acids—H_2S, H_2Se, and H_2Te. The hydrides of all the halogens are acids—HF, HCl, HBr, and HI. As a matter of convention, when the hydrogen is written first in a formula, the compound is an acid; when the hydrogen is written last, the compound is a base. In aqueous solutions, the molecules of HCl, HBr, and HI all completely dissociate into ions, so hydrogen forms hydrogen ions (H^+). (However, because a cation of ordinary hydrogen is just a proton, it is too small and too highly charged just to "float around" in water unattached.

(continued on page 22)

DISASTER IN THE MAKING: THE *HINDENBURG* ZEPPELIN

Gas-filled balloon flight has had a checkered history. Although balloons today are filled with helium, which is not flammable, early gas-filled balloons contained hydrogen, which is highly flammable. In fact, history's first air disaster occurred in 1785 with the explosion of a hydrogen-filled balloon that took the lives of two people.

Commercial balloon flights began through the efforts of the inventor, Count Ferdinand von Zeppelin, born in Germany in 1838, whose work resulted in an airship the size of an ocean liner that could carry paying passengers. Soon, an entire fleet of "zeppelins" was servicing Europe. By the beginning of World War I in 1914, Zeppelin's dirigibles had made almost 1,600 flights and had carried more than 37,000 passengers. Not a single accident had occurred, even though hydrogen was still the gas being used.

After the war, dirigibles were still a popular means of transportation and were far roomier and more comfortable than most of today's jet airliners. Large deposits of helium had been discovered in Texas, so that American dirigibles began to be filled with helium instead of hydrogen, making them much safer than German dirigibles, which still used hydrogen exclusively. In the late 1920s, the leading German airship was the *Graf Zeppelin* (the *"Graf"*), named after Count Zeppelin.

In 1934, a new airship was constructed—the *Hindenburg*—that was so huge, it was in fact almost as long as the RMS *Titanic*. The original plan was to use helium instead of hydrogen—it even installed a smoking room, since the danger of hydrogen gas would not be present. However, because the Nazis had assumed power in Germany, the United States refused to sell helium to Germany. Consequently, the *Hindenburg* continued to use hydrogen. (The smoking room stayed!)

The *Hindenburg*'s first flight was in March 1936. During its short history, the *Hindenburg* made 10 successful round-trip flights to New York and several round-trip flights to Brazil. Then, in May 1937, disaster struck. The *Hindenburg* left Germany and flew to Lakehurst, New Jersey. To the horror of hundreds of spectators watching the *Hindenburg*'s arrival, a burst of flame appeared. Within seconds, the entire airship was on fire, falling to the ground completely ablaze. Of the 97 persons on board, 35 of them died.

For years afterward, people debated the cause of the fire. Did the hydrogen explode? Was there a bomb or other sabotage? The most current opinion is that a spark of static electricity ignited the fabric of the *Hindenburg*'s hull. Once a hole burned through the fabric and hydrogen gas began to escape, the hydrogen itself caught fire, quickly igniting the rest of the hull.

The *Graf* and the *Hindenburg*'s successor, *Graf Zeppelin II*, continued to be flown within Germany for several years, but finally Germany destroyed both airships in 1940 to salvage their aluminum for the war effort. The era of zeppelins was over.

In May 1937, the *Hindenburg* left Germany for Lakehurst, New Jersey. To the horror of spectators watching the *Hindenburg*'s arrival, a burst of flame appeared. Within seconds, the entire airship was on fire, falling to the ground completely ablaze. *(AP Photo/Philadelphia Public Ledger)*

(continued from page 19)

Hydrogen cations attach themselves to water molecules to form hydronium ions, H_3O^+.)

Atoms of metals only form cations. Therefore, when atoms of hydrogen combine with atoms of metals, hydrogen forms "hydride" ions (H^-). Examples are lithium hydride (LiH) and sodium hydride (NaH).

An important property of hydrogen is its ability to form bridges, called *hydrogen bonds,* between two molecules. For example, in a glass of water, the water molecules are attracted to each other by the attraction of a hydrogen atom on one water molecule to the oxygen atom on another water molecule. This attraction, hydrogen bonding, is responsible for many of the properties we associate with water. For such a small molecule, water has an unusually high melting and boiling point. Under the normal atmospheric conditions found on Earth, without hydrogen bonding to hold molecules together more strongly, water would exist only in the gaseous state. Without hydrogen bonding, life on Earth would not exist.

FUEL CELLS: HYDROGEN AND THE ENERGY CRISIS

Concerns about global warming and fears of an oil shortage resulting from recent growth in demand have boosted research into hydrogen fuel cells. Fuel cells can emit zero carbon dioxide—the only by-products being water vapor and heat—and, since hydrogen is the most abundant element in the universe, it seems we should never have a shortage.

Fuel cell technology is not new science. Although it did not produce enough energy to be useful, the first fuel cell was designed and built in 1839 by a Welsh physicist, Sir William Grove, and the technology has been progressing ever since. The basic design is battery-like, with negative and positive terminals (anode and cathode, respectively). At the anode, hydrogen molecules are split by a well-chosen catalyst into protons and electrons. The resulting electron current can provide power to run any electrical device. The protons travel through an electrolyte to reach the cathode, where the protons and electrons combine with oxygen molecules to make water vapor and heat. Alkaline cells, whose electrolytic solutions are sodium hydroxide or potassium hydroxide,

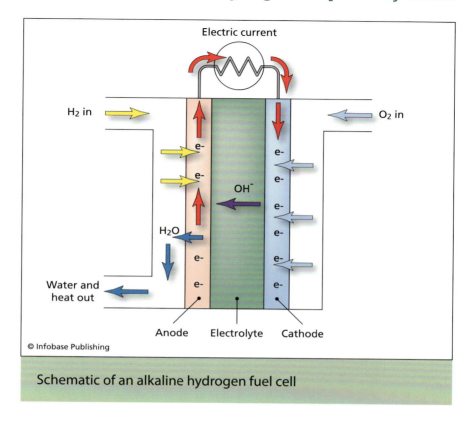

Schematic of an alkaline hydrogen fuel cell

have been used successfully in Apollo spacecraft and the space shuttle as well as small submarines.

Major challenges remain, however, when it comes to developing models suitable for personal automobiles. The abiding problem is cost-effective storage of the hydrogen fuel. Storing it in gaseous form can be ruled out; unless the gas could be highly compressed, which is costly and hazardous, it would need too large a tank. Liquid fuel is a better choice for high-energy density, but its boiling point is −253°C, so it has to be cryogenically cooled. Daimler-Chrysler demonstrated the feasibility of using liquid hydrogen in its NECAR 4 as early as 1999, but cooling and insulation costs made it impractical for mass production.

Current research into metal hydrides, which store hydrogen at a fairly high density (depending on the material), could make fuel tank size practical as well as cost effective. In this type of material, individual hydrogen atoms are absorbed into the lattice of molecules, which,

HYDROGEN AS METAL

As early as 1935, the renowned physicist Eugene Wigner and his colleagues were talking about the possibility of making solid metal hydrogen. Recall that a metal is a material whose atoms or molecules easily share electrons, so a current can flow. This requires a low *resistivity* of the material. The idea was that if hydrogen atoms could be packed together closely enough, the electrons would be as close to one proton as another and be attracted preferentially, depending on small changes in distance. Hence they would move around among the protons.

The density predicted by Wigner was about 25 *Gpa* (or about 250,000 *atmospheres*), but experimentalists have not been successful at making this crystalline form. At Livermore National Laboratory, however, scientists have been able to produce liquid hydrogen with a resistivity of 5×10^{-4} Ω, which is comparable to that of the alkali metals cesium and rubidium at similar temperatures. The pressure required was nearly 10 times higher than predicted, however, and the temperature was twice the predicted value. Such a significant deviation from theory stimulates much interest and discussion in the physics community and is an ongoing topic of research.

The discovery of liquid metallic hydrogen is also significant to planetary scientists. It has long been known that the gas giant Jupiter has a strong magnetic field that cannot be explained by positing an iron core such as that of Earth, as Jupiter's overall density is too low. But its mass is 300 times that of Earth, so the pressure at the center is extremely high—high enough, in fact, to support the hypothesis that liquid metallic hydrogen exists at its core, thus creating the observed magnetic field. Saturn, though not nearly as large, also has pressures high enough for hydrogen metal formation at its core.

The metallic form has only been produced in the lab in a disordered phase. Theorists have speculated about possible production of an ordered state of liquid metallic hydrogen. If created, however, this would be a new and difficult material to classify, as unidentified quantum effects could cause strange phenomena such as multiple phase transitions.

when heated, releases the hydrogen as needed to the anode. Metal hydrides are, however, heavy and difficult to catalyze at normal ambient temperatures.

Intense and competitive research and development in nearly every industrialized country continues, with intriguing results involving lightweight nonmetal hydrides, carbon *nanotubes,* and nanocrystalline metallic alloys. One of the greatest advances may turn out to be the development of miniature fuel cells to power cellular phones, estimated by researchers at Lawrence Berkeley Laboratory to currently use more than 600 million kW-h of energy per year in the United States alone.

TECHNOLOGY AND CURRENT USES

Because it is the most abundant element, hydrogen has naturally become a crucial constituent in myriad processes useful to humankind. Hydrogen is a chemical component of ammonia and hydrochloric acid and consumer products such as soaps and cleaners, as well as cosmetics. For cooking purposes, it converts oils to solid or nearly solid fats, which are more conducive to baking and do not spoil as easily as other fats such as lard. It is used in welding, reduction of metal ores, fixation of nitrogen from the air, *hydrodesulfurization* in natural gas refining, and *hydrodealkylation* to break down aromatic hydrocarbons.

Hydrogen is also important in fuel production. *Hydrocracking* uses the partial pressure of hydrogen gas to break down complex organic molecules, and forms by-products such as ethane, aromatics, and jet fuels. Liquid hydrogen is also used as a rocket fuel. In fission-based nuclear reactors, heavy water (where deuterium replaces regular hydrogen) is used as a neutron moderator.

Fusion research currently relies on the deuterium-tritium fusion reaction to produce useable power, a method that is well understood but still highly inefficient—much more energy must be put into the process than is produced. Fusion is not expected to be a viable source of power for humankind for at least the next 50 years.

For powering automobiles, hydrogen fuel cells may someday be commonly used. The basic design is battery-like with negative and positive terminals (anode and cathode respectively). At the anode, hydrogen molecules are split by a well-chosen catalyst into protons and electrons. The resulting electron current can provide power to run any electrical device. The protons travel through an electrolyte to reach the cathode, where the protons and electrons combine with oxygen molecules to make water vapor and heat. Major challenges remain, however, when it comes to developing models suitable for personal automobiles, the most difficult being cost-effective storage of the hydrogen fuel. The slow pace of development in this area is the reason President Obama in May 2009 reduced federal funds for hydrogen fuel research by more than half.

One of the most marketable of hydrogen products is hydrogen peroxide (H_2O_2), with applications ranging from household antiseptic to rocket propellant. Hydrogen is also an important element in all biological compounds. Clearly this element will continue to play a major role in research in every scientific discipline.

2

Carbon: The Element of Life, Coal, and Diamonds

Carbon—element number 6—is the most versatile element in the periodic table. The singular ability of carbon atoms to form an incredibly diverse number of compounds is due to the ability of carbon atoms to form strong covalent bonds with other carbon atoms and to do so such that long chains and rings of atoms can form. Because the molecules that comprise living organisms—sugars, starches, fats, oils, *proteins,* and *nucleic acids*—are based on carbon, carbon is the "element of life," and substances that contain at least carbon and hydrogen are called *organic compounds.* In fact, scientists do not know of any living organisms that do not contain carbon. In addition to the molecules of life, carbon has an important role in *inorganic* chemistry. Coal and diamond both represent elemental forms (allotropes) of carbon, and carbon dioxide is one of the most important gases in Earth's atmosphere.

27

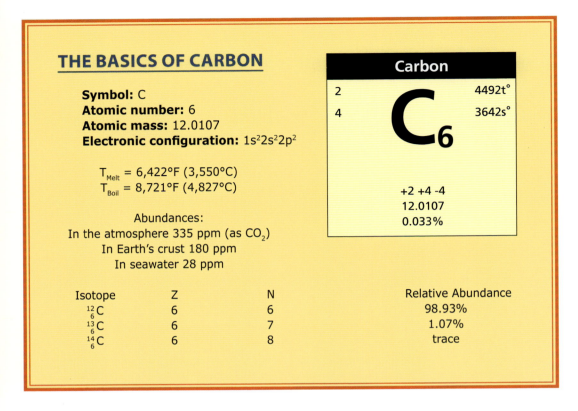

THE BASICS OF CARBON

Symbol: C
Atomic number: 6
Atomic mass: 12.0107
Electronic configuration: $1s^2 2s^2 2p^2$

T_{Melt} = 6,422°F (3,550°C)
T_{Boil} = 8,721°F (4,827°C)

Abundances:
In the atmosphere 335 ppm (as CO_2)
In Earth's crust 180 ppm
In seawater 28 ppm

Carbon			
2			4492t°
4	C$_6$		3642s°
	+2 +4 -4		
	12.0107		
	0.033%		

Isotope	Z	N	Relative Abundance
$^{12}_{6}C$	6	6	98.93%
$^{13}_{6}C$	6	7	1.07%
$^{14}_{6}C$	6	8	trace

THE ASTROPHYSICS OF CARBON

A young star fuses hydrogen into helium—a process called "hydrogen burning"—in its core until the only hydrogen left is in the outer shell of the star. At that time, the helium core begins to collapse under its own weight while hydrogen burning continues in the outer shell. Enormous amounts of heat and light are generated by the energy of the collapse, generating radiation pressure that makes the outer shell expand explosively while the core continues to contract. This is the red-giant phase of a star, so called because the expanding gas in the outer shell filters out all but the star's red wavelengths.

The collapsing core meanwhile becomes denser and hotter until helium is able to fuse into carbon via the *triple-alpha process* (helium burning). Initially two helium nuclei, or alpha particles, fuse to form an unstable and very short-lived state of beryllium. Despite the brevity of its existence (about a tenth of a *femtosecond*), there is a chance for some of these [8]Be nuclei to capture alpha particles to form ^{12}C, the basis of life as it is known. The star now begins to form a central core of carbon. These reactions can be written as the following equations:

$$_2^4\text{He} + {}_2^4\text{He} \rightarrow {}_4^8\text{Be}$$

$$_4^8\text{Be} + {}_2^4\text{He} \rightarrow {}_4^{12}\text{C} + \gamma.$$

where "γ" is a high-energy photon or "gamma ray."

The helium burning in turn produces radiation pressure around the core, so there is no further collapse until all the helium is used up. The conditions are also suitable for some carbon atoms to capture alpha particles to produce ^{16}O, so the red-giant stellar phase comprises helium, carbon, and oxygen nuclei. The carbon/oxygen ratio is of compelling interest in astrophysics because the subsequent heavy-ion burning stages rely heavily on this relative abundance. A clearer understanding of the alpha-capture process at this stage of stellar development would advance knowledge of what percentage of stars should progress to white dwarfs, red supergiants, or supernovae.

After all the helium is consumed at the core, without radiation pressure for support, a star will once again begin to collapse under its own weight. At sufficient pressure and temperature, carbon burning can take place, but only if the star's weight is enough to effect extreme density (2×10^8 kg/m³). For stars at least four times as massive as the Sun, carbon fusions produce—with varying probabilities—neon, sodium, and magnesium.

EARTHBOUND: FROM COAL TO DIAMONDS

One of the most common elements on Earth, carbon comes in diverse forms, including graphite, amorphous carbon, diamond, and fullerene. The different forms, or *allotropes,* are distinguished by different bonding modes. All carbon deposits developed over millions of years from compaction of formerly living plant and animal cells. For that reason, coal and some of its by-products have earned the name *fossil fuels.*

Graphite, the purest form of coal, has been mined in this form in only one location—a large deposit in Cumbria, England. It is usually found with an admixture of other minerals such as quartz or mica, with the largest producers being China and India. Madagascar has become an important producer of large-flake graphite, some with an intriguing rhombohedral phase. The mining there, however, is increasingly becoming a threat to a unique biodiversity.

High-grade common coal, often extracted from anthracite, began its formation in swamp and wetland systems where oxidation rates were low, allowing large *sedimentary* and *metamorphic* deposits to form worldwide. Coal is easily extractable and easily transported, making it a popular choice for use as a fuel in electric power plants. The risks associated with mining procedures, however, have earned it the reputation as the deadliest source of power in history, and coal power plant emissions are a serious global warming hazard.

The hardest of carbon's allotropes is the coveted and rare diamond, found famously in Africa—particularly South Africa, Botswana, and Angola—but also in Russia, India, Australia, and Canada. The United States boasts only one source of this gemstone: Crater of Diamonds State Park in Arkansas. Diamond mines are generally situated near a volcanic pipe, as the densely packed carbon, formed under extreme pressure in the depths of the Earth (more than 93 miles [ca. 150 km] beneath the surface), must be brought up by volcanic activity. Though fairly brittle, diamond is ideal for use in jewelry because of its high *index of refraction,* which allows for bending of light at angles unreachable by other stones. (Glass has an index of refraction about two-thirds that of diamond.)

Fullerene, a hollow-core allotrope of carbon, was first seen in 1992 in a sample of metamorphic rock from northwestern Russia, though it had been produced a few years earlier in university laboratories. Configurations range from soccer-ball shapes and *icosahedrons* to nanotubes and nanofibers.

Other recently discovered allotropes include a *nanofoam*—a web of light magnetic carbon clusters, lonsdaleite—a sort of disfigured hexagonal diamond lattice, and an aggregated diamond nanorod.

DISCOVERY AND NAMING OF CARBON

Coal, charcoal, graphite, and diamonds were all known in prehistoric times, but they probably were not recognized as all being forms of the same element—carbon. Ancient civilizations were most likely to use

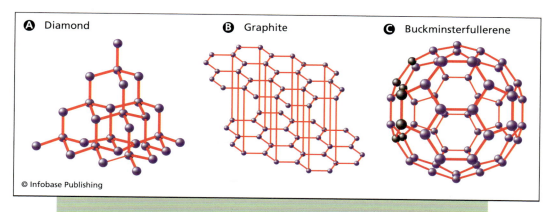

The structures of (A) diamond, (B) graphite, and (C) buckminster-fullerene.

charcoal as a source of fuel, and less likely to use carbon in its other forms. Even today, charcoal is a common fuel in various parts of the world. The name itself—*carbon*—is derived from *carbo*, the Latin word for charcoal.

It was not until the beginning of modern chemistry that chemists began to recognize carbon's varied forms. The biggest dilemma occurred with diamond, since its appearance is significantly different from the appearances of coal, charcoal, or graphite, and diamond is exceptionally harder than the others are. In 1694, chemists discovered that sunlight focused in by a large magnifying glass causes a diamond to disappear. In 1771, a diamond was heated and shown to burn completely without leaving any ash. It was not until 1796, however, that Smithson Tennant, an English chemist, demonstrated that when a diamond was burned, the only product was carbon dioxide, indicating that the diamond itself was a form of carbon.

Carbon mainly is found in the form of *hydrocarbons*—natural gas, oil, and coal—and carbonate-containing minerals such as limestone ($CaCO_3$). Carbon also exists in the form of carbon dioxide, which makes up 0.0335 percent of Earth's atmosphere.

There are three naturally occurring isotopes of carbon: carbon 12, which makes up 98.9 percent of all the carbon on Earth, carbon 13, which makes up 1.1 percent of the carbon, and carbon 14, which is radioactive

Coal is usually found with an admixture of other minerals, such as quartz or mica, though it seldom resembles diamond. The largest producers in the world are China and India. (*Pixelmaniak/iStockphoto*)

(half-life = 5,730 years) and exists in only trace amounts. Carbon 11 can be produced artificially. Although it is radioactive with a half-life of only 20 minutes, it is an effective agent of medical diagnosis.

Another class of carbon allotropes was discovered in 1985 by Harold W. Kroto, James R. Heath, Sean O'Brien, Robert Curl, and Richard Smalley. Soccer-ball-shaped spheres of 60 carbon atoms with formulas like C_{60} and C_{70} were found in carbon soot and later recognized to be ubiquitous in interstellar clouds. C_{60} is recognized as the most perfectly spherical known molecule. Because the arrangements of the carbon atoms resemble the architecture of geodesic domes, which were invented by Richard Buckminster Fuller, this class of carbon allotropes came to be called *fullerenes*. Kroto, Curl, and Smalley shared the 1996 Nobel Prize in chemistry for this discovery.

A newly discovered form of carbon called *graphene* exhibits high thermal conductivity and an unprecedented electron mobility: Electrons in graphene move practically uninhibited by the atomic lattice. First symbolized in 2004 by researchers at the University of Manchester in England, this material—whose configuration resembles chicken wire, as would a single layer of graphite—is also extremely strong. Obvious applications in electronic circuits, which rely on electron mobility for signal speed, have made the study of graphene priority research in universities around the world.

THE CHEMISTRY OF CARBON

Chemical compounds are divided into two major groups: organic compounds and *inorganic* compounds, classifications assigned by the Swedish chemist Jöns Jakob Berzelius in 1807. Organic compounds were said to be those substances that are derived from living organisms—plants or animals. All other substances were said to be inorganic and would be derived from minerals. This distinction between organic and inorganic

The 1996 Nobel Prize in chemistry was awarded to Robert Curl (center), Richard Smalley (right), and Sir Harold Kroto (left) for their discovery of carbon fullerenes. (*AP Photo/Soren Andersson*)

substances seemed reasonable at the time. It was supported by the general observations that organic substances can be converted fairly easily into inorganic substances (such as carbon dioxide and water), but no one had observed the conversion of inorganic substances into organic substances.

The situation changed in 1827, when German chemist Friedrich Wöhler synthesized urea, one of the body's waste products in urine, by heating ammonium cyanate, an inorganic compound. At first Wöhler could not believe the results. He repeated the experiment several times just to make sure he had not made a mistake. Finally, in 1828, he published his findings and surprised the chemical world.

Today chemists no longer classify organic versus inorganic compounds based on how they are derived. The current textbook definition of an organic compound is a pure substance that contains carbon and hydrogen and possibly other elements. In contrast, inorganic compounds are all the remaining pure substances (other than simple elements) that occur either naturally or artificially.

Carbon atoms readily form covalent bonds with other carbon atoms and with atoms of other nonmetals, especially hydrogen, nitrogen, oxygen, phosphorus, sulfur, and the halogens. Carbon atoms form these bonds by sharing pairs of electrons with atoms of other elements. When two atoms share two electrons, the bond is called a single bond (symbolized in a structural formula by a single dash "–"). When four electrons are shared, the bond is called a *double bond* (symbolized by a double dash "="). When six electrons are shared, the bond is called a *triple bond* (symbolized by a triple dash "<≡>"). A carbon atom will

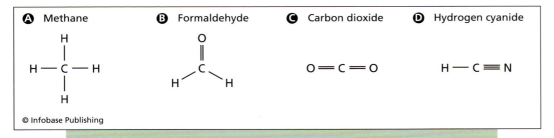

Different types of chemical bonding in carbon

form enough bonds with other atoms so that a total of eight electrons is almost always shared.

To share a total of eight electrons with other atoms, a carbon atom may exhibit more than one type of bonding. The possibilities are the following: (1) form four single bonds, as in methane, CH_4; (2) form a double bond and two single bonds, as in formaldehyde, H_2CO; (3) form two double bonds, as in carbon dioxide, CO_2; or (4) form a triple bond and a single bond, as in hydrogen cyanide, HCN.

When carbon atoms bond primarily to atoms like hydrogen or oxygen, small molecules like methane (CH_4) or carbon dioxide (CO_2) are possible. Alternatively, when carbon atoms bond to other carbon atoms, long chains of atoms are possible. Carbon chains also sometimes loop back on themselves and form rings. Different groups of carbon compounds are classified according to their structures as chains, as rings, or as compounds containing other elements besides carbon and hydrogen—oxygen and nitrogen being the two most important other elements that may be part of organic compounds. What kinds of bonds form will determine the *functional group* to which that compound belongs and thus determine the compound's chemical and physical properties. Examples of different kinds of bonds and the functional groups they represent are given in the following table.

Hydrocarbons are classified according to whether the carbon atoms are linked by all single bonds (the *alkanes*), or whether a double bond (the *alkenes*) or triple bond (the *alkynes*) is present. Chains can also be classified as *straight chains,* in which no branching occurs, or as *branched chains,* in which there are side chains to the main chain of carbon atoms.

Hydrocarbons and related compounds can also exist in rings, in which case they are called *cyclic compounds.* An important example of a cyclic compound is benzene. The benzene ring is relatively stable, making benzene fairly nonreactive. Benzene is an important solvent in the chemical industry. Compounds derived from benzene are used as gasoline additives to boost performance.

Synthetic plastics, fibers, and *elastomers* (rubber) are major products of the chemical industry. All of these substances consist of very large carbon chains called *macromolecules* or *polymers,* which may have

COMMON FUNCTIONAL GROUPS OF CARBON

GROUP	ELEMENTS	BONDS	EXAMPLE
Alkanes	C, H	all single	C_8H_{18} Octane, a component of gasoline
Alkenes	C, H	C = C double bond	$H_2C = CH_2$ Ethylene, a raw ingredient for common plastics
Alkynes	C, H	C ≡ C triple bond	HC = Acetylene, used in welding
Alcohols	C, H, O	all single	$H_3C – CH_2OH$ Ethanol, alcoholic beverages
Ethers	C, H, O	all single	$H_3C – CH_2 – O – CH_2 – CH_3$ Diethyl ether, anesthetic
Aldehydes	C, H, O	C = O double bond	$H_2C = O$ Formaldehyde, preservative
Ketones	C, H, O	C = O double bond	$$\begin{matrix} & O \\ & \| \\ H_3C – & C – CH_3 \end{matrix}$$ Acetone, paint thinner
Carboxylic acids	C, H, O	C = O double bond and C – O single bond	$$\begin{matrix} H & & O \\ \| & & /\!/ \\ H – C & – & C \\ \| & & \| \\ H & & O – H \end{matrix}$$ Acetic acid, component of vinegar
Amines	C, H, N	all single	$CH_3 – NH – CH_3$ Dimethyl amine, tanning agent

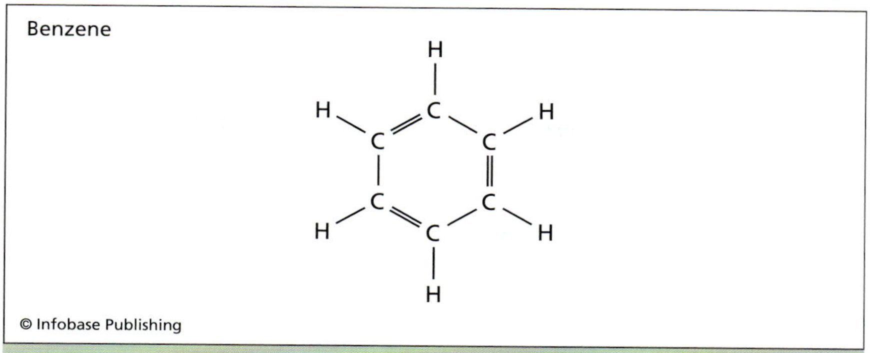

Benzene

© Infobase Publishing

The structure of benzene

hundreds, or even thousands, of carbon atoms in them. Most synthetic polymers are manufactured from the raw ingredients found in natural gas or petroleum.

Molecules of *alcohols* and *ethers* contain an oxygen atom; both kinds of compounds have the general molecular formula $C_nH_{2n+2}O$. (For example, if n = 4, then the formula would be $C_4H_{10}O$.) Therefore, alcohols and ethers containing the same number of carbon atoms are *isomers* of each other, even though they have significantly different properties. Ethanol and dimethyl ether both have the molecular formula C_2H_6O but an important difference in their structural formulas.

The most common ether is diethyl ether. Because of diethyl ether's *volatility*, it is commonly used as a starting fluid in motor vehicles. In the 19th century, diethyl ether was used as an *anesthetic*. Its low *flash point*, however, made it hazardous to use in the presence of oxygen and any flames or other ignition sources.

Ethanol is typically made from plant *sugars*, such as the glucose or maltose found in grains, through the process of *fermentation* in which the following reaction occurs in the presence of yeast:

$$C_6H_{12}O_6 \text{ (s)} \rightarrow 2\ C_2H_5OH \text{ (}l\text{)} + 2\ CO_2 \text{ (g)}.$$

Ethanol is the alcohol present in all alcoholic beverages. (Any other alcohol would be far too toxic for human consumption. Even ethanol is toxic, hence the term *intoxicated.*)

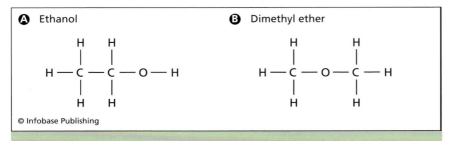

The structures of ethanol and dimethyl ether

Molecules of *aldehydes* and *ketones* also contain an oxygen atom, but in this case, the oxygen atom and a carbon atom are connected together with a double bond (see table on page 36). A carbon-oxygen double bond is depicted in a structural formula as "C = O." The general formula for an aldehyde or ketone is $C_nH_{2n}O$. An aldehyde has its oxygen atom attached to a carbon atom at the end of the hydrocarbon chain, and a ketone has its oxygen atom attached to a carbon atom that is not at the end of the chain. Both compounds have good solvent properties.

Esters and *carboxylic acids* contain two oxygen atoms—one with a single bond to a carbon atom, and the other with a double bond to the same carbon atom (see figure on page 39). Many esters occur naturally, are good *nonpolar* solvents, and tend to be fragrant, for which purpose they are often used in household products. Ethyl acetate, for example, is the solvent often used in fingernail polish remover.

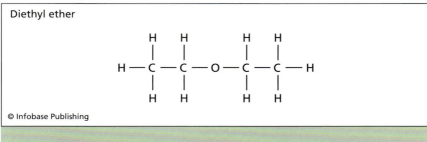

The structure of diethyl ether

Ethyl acetate

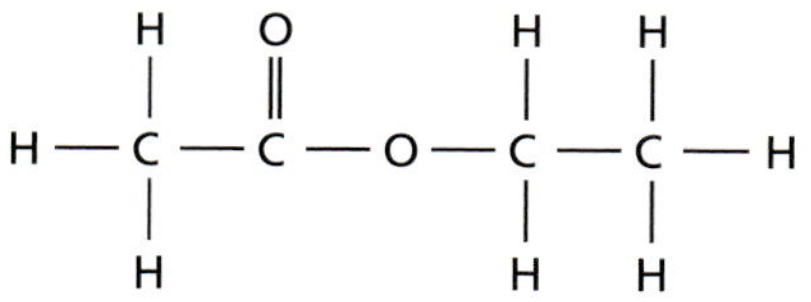

$$H-\overset{\overset{\displaystyle H}{|}}{\underset{\underset{\displaystyle H}{|}}{C}}-\overset{\overset{\displaystyle O}{\|}}{C}-O-\overset{\overset{\displaystyle H}{|}}{\underset{\underset{\displaystyle H}{|}}{C}}-\overset{\overset{\displaystyle H}{|}}{\underset{\underset{\displaystyle H}{|}}{C}}-H$$

© Infobase Publishing

The structure of ethyl acetate

Carboxylic acids occur naturally, as the following list shows:

Name of Compound	Where Found
formic acid	the sting of ants
acetic acid	vinegar
butanoic acid	rancid butter
caproic acid	goats milk
oxalic acid	rhubarb

Because carboxylic acid molecules contain an –OH group, the molecules can hydrogen-bond to each other, resulting in relatively high boiling points. They can also hydrogen-bond to water molecules resulting in *miscibility* with water.

Amines contain nitrogen atoms (see table on page 40) and are among the most abundant organic molecules found in nature. In contrast to carboxylic acids, which are nature's weak acids, amines are nature's weak bases and are similar to ammonia (NH_3) in that sense. Many synthetic organic compounds are amines or contain nitrogen groups. Naturally occurring compounds include amino acids, peptides, proteins, alkaloids, and neurotransmitters. Synthetic compounds include decongestants, anesthetics, sedatives, and stimulants.

Because amines contain an –NH group, they exhibit hydrogen bonding, although to a lesser extent than alcohols do. Because of hydrogen bonding, they have relatively high boiling points, and small amines are *miscible* with water. The follow list shows some of the amines that are physiologically active.

Name of Compound	Application
epinephrine (adrenaline)	stimulant
propylhexedrine	nasal decongestant
amphetamine	antidepressant
mescaline	hallucinogen
caffeine	stimulant

Sometimes amines have rather creative names that are suggestive of their odors: It should not be difficult to imagine what *cadaverine* and *putrescine* smell like!

THE BASIS OF LIFE

Organic compounds are the building blocks of living organisms. Carbohydrates and *lipids* are composed of carbon, hydrogen, and oxygen. *Carbohydrates* are grouped as sugars, starches, or cellulose. Sugars and starches in our foodstuffs become the substances that fuel our metabolic processes. Cellulose, which forms the cell-wall material in plants, is not digestible by humans. Fats, oils, and *steroids* are examples of the lipids found in our foods. Lipids also fuel our metabolism, are important components of our cell membranes, and include substances like cholesterol, cortisone, and estrogen that are essential to healthy bodies. Proteins are composed of carbon, hydrogen, oxygen, nitrogen, and sometimes sulfur. The functions of proteins are extremely diverse, and include catalytic activity, transport and storage mechanisms, and structural importance. The two nucleic acids are DNA and RNA, which are responsible for the storage and replication of our genetic code.

Examples of common simple sugars are glucose, fructose, and lactose. Starches and cellulose are polymers made by linking sugar molecules together in long chains. Foods described as *complex carbohydrates* contain starches. Digestive *enzymes* (which themselves are a kind of protein molecule) in saliva and in stomach juices catalytically break starch molecules into their component sugar molecules.

Fats contain hydrocarbon chains that are completely *saturated,* meaning that they have all carbon-carbon single bonds and contain the maximum number of hydrogen atoms they can hold. In oils the chains tend to be partially *unsaturated,* meaning there are one or more carbon-carbon double bonds. Fats, like lard, tend to be solid or semisolid at room temperature and are derived from consumption of animal products. Oils, like vegetable oils, tend to be liquid at room temperature and are derived mostly from plant products. Steroids contain *fused rings* of carbon, hydrogen, and oxygen atoms. Humans can assimilate steroids from the foods they eat. Alternatively, the bodies of humans can also manufacture the steroids we need.

Proteins are polymers of amino acids. An amino acid has a backbone consisting of a carbon atom with linkages to a carboxylic acid group, an amine group, a hydrogen atom, and a group of atoms called a *residue.* There are 20 naturally occurring amino acids. Important proteins include enzymes, which are biological catalysts; hemoglobin, which transports oxygen through the blood stream; keratin, which makes up hair, nails, and feathers; and myosin, which makes up muscles.

DNA, or deoxyribonucleic acid, is the "molecule of life." DNA serves as a template for the body's syntheses of protein molecules. Embedded in the genetic code of the *genes* and *chromosomes* that make up DNA are the instructions necessary to make every protein our bodies need. Molecules of RNA resemble DNA. RNA picks up the information from DNA and carries that information to the cell's *cytoplasm,* where protein synthesis occurs from the mix of amino acids already present in the cytoplasm.

One of the important puzzles in answering the question "How did life begin?" revolves around the enzyme molecules in all living organisms. DNA is necessary to make enzymes; enzymes are necessary to make DNA. So which came first? And how could life have started without both kinds of molecules present?

PETROLEUM DEPOSITS AND OIL DEPLETION

Ever since the origin of life on Earth approximately 3 billion years ago, carbon compounds have been accumulating on the ocean floors as generations of organisms expired and drifted down to form layer upon layer of rich sediment. Some of that sediment, when compressed at just the right temperature and depth, transformed into the hydrocarbon substance we now call oil. Trapped between layers of less porous rock, huge deposits of the black viscous substance lay undisturbed for eons far beneath Earth's surface.

At even greater depths, the carbon compounds were subjected to higher temperatures and hence tended toward a gaseous form, termed *natural gas.* Methane (CH_4) is the most common form, though ethane, propane, and butane make up about 20 percent of deposits.

Although humans as far back as 4000 B.C.E. used oil from surface seepage, it was not until the growing population needed oil for kerosene lamps and automobiles that it was actively extracted from underground. Since the early 1900s, more than a trillion barrels of oil have been pumped worldwide. Estimates place reserves at about another trillion barrels (42 trillion gallons). Only one large oil field has been discovered in the last three decades—in a brutal offshore climate in the north Caspian Sea. Worldwide oil production has most likely reached its peak, but demand is steadily growing.

Peak oil production is an idea that was first introduced in the 1950s by Marion King Hubbert, a geophysicist then working for the Shell Oil Company. Hubbert was interested in rates of production and depletion of any nonrenewable resource. Taking into account the timing of discovery of new wells and the rate of pumping, he forecast that peak production of oil in the United States would occur right around the year 1970. His prediction was the target of much skepticism from his contemporaries, but he turned out to be correct. Implementing the same concepts as Hubbert, geophysicists have estimated that world oil production should peak around 2010, with a margin of error between five and 10 years. In fact, as of this writing, production by the world's three largest oil fields is reported to be in decline by rates exceeding 7 percent per annum.

An intimately linked problem is that world demand for oil is growing at an average rate of a little more than 2 percent per year. At this growth rate, demand will double in less than 35 years. The growth in demand coupled with decreasing oil production presents a distinct problem, most notably in the transportation industry. In the United States, automobile manufacturers are already in deep economic trouble because of their concentration on SUVs and other large

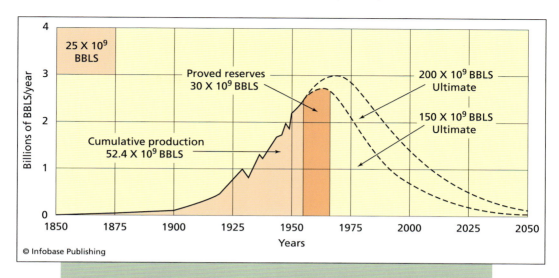

M. King Hubbert's original prediction for "ultimate United States crude oil production based on assumed initial reserves of 150 and 200 billion barrels" (Source: "Nuclear Energy and the Fossil Fuels," *Drilling and Production Practice,* Publ. no. 95, American Petroleum Institute. June 1956)

vehicle production, and there is growing concern about dependence on foreign oil.

As a result, hybrid vehicles like the Toyota Prius that run on a combination of internal combustion and battery power are growing in popularity, as is biofuel investigation. Research into hydrogen-fueled vehicles, however, had its funding cut in 2009 by the Department of Energy due to the dubious economic feasibility of this technology. All-electric vehicles are finally receiving some attention: Nissan released the Hyper Mini in 2000 and may have a new breakthrough with the Pivo concept car, which uses a compact lithium ion battery. Tesla Motors unveiled its lithium-ion battery propelled sports car, the Tesla Roadster, in 2006 and began delivery to buyers in 2008. This all-electric vehicle, currently built in England, can run up to 240 miles (386.2 km) on a single charge. Even with a price tag of more than $100,000 by September 2009 the company had sold 700 roadsters. The cost makes this environmentally friendly vehicle unattainable for most consumers, however.

Carbon dioxide is removed from the atmosphere by photosynthesis in green plants and solution in aquatic systems like this marsh; its average residence time in the atmosphere is about 100 years. *(Jaap Hart/iStockphoto)*

THE CARBON CYCLE

Carbon dioxide (CO_2) comprises 0.03 percent of the atmosphere. Carbon dioxide is removed from the atmosphere by *photosynthesis* in green plants and solution in aquatic systems; its average *residence time* in the atmosphere is about four years. Photosynthesis, respiration by living organisms, and their decomposition represent the major pathways in carbon's cycle through the *biosphere.* Through photosynthesis, carbon becomes incorporated into the tissues of living organisms. The amount of carbon stored in living organisms (*biomass*) is about the same as the amount of carbon present in the atmosphere. When CO_2 dissolves in freshwater or saltwater environments, it is converted into bicarbonate (HCO_3^-) and carbonate (CO_3^{2-}) ions. These negatively charged ions can combine with positively charged ions like calcium (Ca^{2+}) to form minerals (e.g., limestone, $CaCO_3$) or the shells of mollusks (e.g., snails and

CARBON DATING

Because all living things take in carbon from the air, either directly or indirectly, and intake stops when they die, scientists are able to gauge the approximate age of bones, fossils, wood, or anything that once lived. The method is based on the changing ratio of carbon isotopes in the material.

While living, plants absorb CO_2 from the air. Some of the carbon exists in the form of the *radioactive* isotope ^{14}C, which forms as *cosmic rays* (mostly protons) continuously arriving from space interact with the upper atmosphere. Commonly, the protons interact with oxygen atoms to produce neutrons. Those neutrons can then interact with nitrogen in the atmosphere to form ^{14}C atoms. The following equations illustrate these reactions:

$$^{17}_{8}O + ^{1}_{1}P \rightarrow ^{17}_{9}F + ^{1}_{0}n$$

$$^{14}_{7}N + ^{1}_{0}n \rightarrow ^{14}_{6}C$$

$$^{14}_{6}C \rightarrow ^{14}_{7}N + ^{-1}_{0}e$$

Animals absorb carbon by ingesting plants or other animals. The ratio of ^{12}C—the most abundant stable isotope of carbon—to ^{14}C in a living system remains constant. Upon death, however, ingestion stops and the ^{14}C, which is always undergoing radioactive decay, is no longer replaced in the cells.

As the ^{14}C continues to decay with a *half-life* of 5,568 years, the isotopic ratio changes accordingly, and can thus be used to monitor the date of origin of the sample in question. The method is valid for determining with an error of only about 30 years the age of an object younger than 50,000 years. Other methods, such as thorium- and potassium-dating, are available for dating much older samples.

clams). In turn, through respiration and other biochemical processes, living organisms return carbon dioxide to the atmosphere. Weathering and erosion processes break down rocks and recycle their carbon. By far the greatest quantity of carbon is stored in sedimentary rocks.

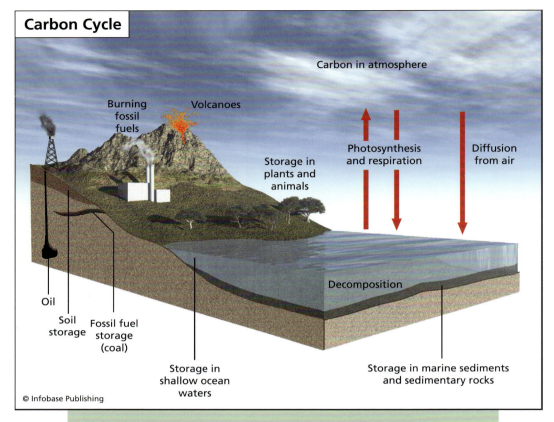

Carbon Cycle

Carbon in atmosphere

Burning fossil fuels

Volcanoes

Photosynthesis and respiration

Diffusion from air

Storage in plants and animals

Oil

Soil storage

Fossil fuel storage (coal)

Storage in shallow ocean waters

Decomposition

Storage in marine sediments and sedimentary rocks

© Infobase Publishing

The carbon cycle

The products of plant photosynthesis sequester carbon. *Sequestration* can be long term or short term; the overall average residence time in terrestrial plants is about 16 years and only a few months in marine plants. In long-term storage, the plant materials are buried underground and gradually become fossil fuels—coal, petroleum, natural gas—where the carbon in them may be sequestered for millions of year. (See the "Global Warming and CO_2" section of this chapter.)

In short-term storage, carbon is incorporated into agricultural or other food products whose lifetimes are measured in weeks or months or into forest products whose lifetimes are measured in decades or centuries. When food products are consumed, some of the carbon is stored in the bodies of the *herbivores* that have eaten the plants and some of the carbon is returned to the atmosphere as CO_2. Herbivores, in turn, may be eaten by *carnivores,* and the carbon is then stored in their bodies. Eventually, all plants and animals die, and *decomposers* like bacteria and fungi recycle the carbon and other minerals found in decaying organisms. When forest products are harvested for wood products or paper pulp, the carbon in them is stored for periods ranging from days to centuries, after which the carbon in them, too, will finally be recycled.

GLOBAL WARMING AND CO$_2$

Since the beginning of the industrial revolution, the amount of carbon dioxide (CO_2) in the atmosphere has been steadily climbing. While CO_2 fluctuations, mainly due to the random nature of volcanic eruptions, have been common throughout Earth's history, the increase in the past 200 years is a direct result of the burning of fossil fuels by humans. There was very little consumption of fossil fuels until coal began to be utilized during the 18th century and, more recently, petroleum and natural gas began to be utilized during the 20th century. Since World War II, the rate of consumption of fossil fuels has increased exponentially so that millions of years worth of stored carbon is now being returned to the atmosphere (as CO_2) at a much faster rate than it is being removed from the atmosphere. Consequently, the concentration of CO_2 in Earth's atmosphere is increasing at an alarming rate, with potentially serious environmental consequences.

For most of recorded history, humans derived power from regional sources for pulling plows, turning shafts, and cooking. Draft animals provided horsepower. Streams turned water wheels, and local wood or peat was burned for heating and cooking. In the early 1800s, however, whale oil became the lamp fuel of choice for those who could afford it, and for the first time massive amounts of fuel began to be shipped over long distances. Whale oil was so popular that the whale population declined severely, nearly driving several species to extinction. But burning animal fat had its downsides—it was costly, smoky, and had a short shelf life and an offensive smell. When a clean-burning inexpensive kerosene was developed in 1857, it quickly became the exclusive fuel for lighting streets and homes.

Around the same time, human population growth began an unprecedented climb, mainly due to improved health care for children, which rapidly increased the need for transportation of food and other goods

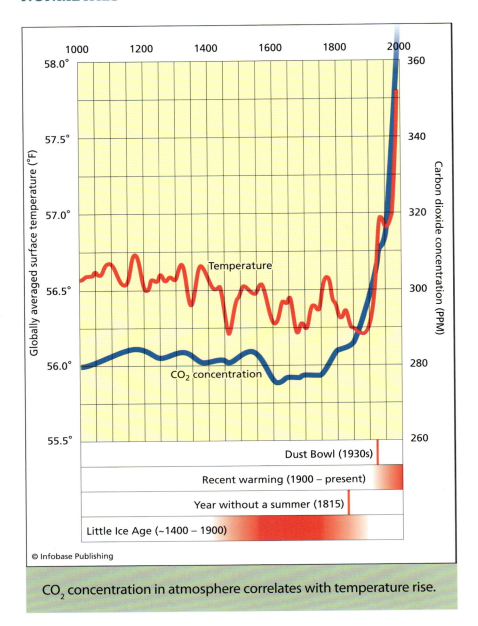

CO$_2$ concentration in atmosphere correlates with temperature rise.

over large distances. The ensuing development of the coal-fired steam engine and especially the internal combustion engine led to a boom in the manufacture of consumer goods. Both allowed speedy transport of materials and people.

Concurrently, the innovative efforts of inventors like Thomas Edison and Nikola Tesla resulted in the ability to provide every household in the country with electricity to power lights, stoves, and other conveniences. Centralized power plants, most of them coal-burning, began to pop up near the larger population centers. Demand has fueled growth ever since. By conservative estimates, coal-fired power plants now emit 2.204 billion tons (2 billion *tonnes*) of CO_2 per year in the United States alone. Internal combustion automobiles contribute another 346 million tons (314 million tonnes) annually (2004 figures).

Additional CO_2 increases the fraction of solar energy retained by the planet's system. Carbon dioxide and other greenhouse gases (GHGs) such as methane, ozone, nitrous oxide, and water vapor preferentially absorb and re-emit infrared radiation—heat—restricting its

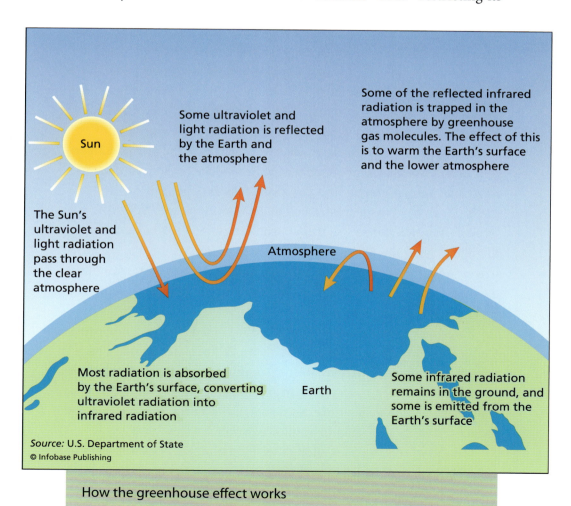

Source: U.S. Department of State
© Infobase Publishing

How the greenhouse effect works

normal escape from the atmosphere. The more greenhouse gas in the system, the more heat is trapped. This is known as the "greenhouse effect." Although CO_2 is by far the largest contributor to the problem, other GHGs, largely associated with farming and ranching, are raising concerns. Another carbon molecule, methane (CH_4), which is produced by livestock manure and decaying vegetable matter as well as burning of trees and grasses, has reached its highest concentration in the atmosphere in at least 400,000 years. Short-lived compared with an average CO_2 molecule, a molecule of methane is able to trap 20 times the heat. While recent droughts have slowed the decay of some areas of vegetation, the recently observed melting of huge areas of permafrost in Siberia has the potential to release billions of tons of the gas in the near future. Since the end of the last ice age, methane has constantly been released by the underlying peat bogs as they decayed. The gas could not escape into the air, but formed bubbles in the permafrost, effectively trapping incalculable volumes of methane. Scientists project that, over the next decade, continued melting of permafrost in Siberia and other areas could significantly accelerate global atmospheric warming.

Recognizing the risks associated with warming the planet, 62 percent of the world's CO_2-producing nations worked out a plan in 1997 to reduce annual greenhouse gas emissions to 5 percent below 1990 levels. Much hard work and trained negotiation went into achieving the level of international cooperation needed to get 169 nations to agree to the plan, called the Kyoto Protocol in recognition of the city in Japan where the gathering was held. Australia and the United States famously decided to opt out of the agreement, their major objection being the perceived cost of implementation.

Based on the guidelines put forth by the United Nations Framework Convention on Climate Change, the Kyoto Protocol designated a target year of 2012 for countries to meet the reduction goal. To achieve this, each country agreed to an effort that seemed logistically and economically viable in the near future. Some agreed to more than a 5 percent reduction, while some had to be allowed an increase.

On the whole, the plan, though well intentioned, does not seem to be working. Japan, which aimed for a decrease of 6 percent from 1990 levels, reports an increase of 8 percent in annual emissions, according to 2007 data; Spain's increase from 2003 to 2004 was 4.8 percent; and

Canada's pledge has failed miserably. The European Union is having better success at controlling output, but recent measurements display an upward rather than downward trend in emissions. The difficulties with compliance seem to be largely attributable to public and private misconceptions about the actual risks from excess GHGs and the costs of implementing solutions. Preoccupation with local economic concerns and maintaining a modern lifestyle has left little room for consideration of global problems.

To address the question of the current and future economic state of affairs in this regard, economist Nicholas Stern at the request of the British government, compiled a 700-page report on predicted economic effects of global climate change in October 2006. While controversial, the Stern report was well informed and concise, stating in clear terms that a 1 percent annual investment in carbon reduction could alleviate serious economic impacts of climate change; the report also warned that ignoring the issue would have dire consequences.

The same year, a powerful documentary entitled *An Inconvenient Truth*, envisioned by and starring former U.S. vice president Al Gore, attempted to awaken the general public as well as world governments to the correlation between increased CO_2 in the atmosphere and rise in temperature and to the ensuing risks to Earth's ecological balance. Wide acclaim resulted, and the film received two Academy Awards, including one for Best Documentary Feature Film. The success of the film may well have been a catalyst for subsequent political, industrial, and social activity around a push for responsibility in decreasing GHG emissions worldwide. Indeed, the Nobel Prize committee believed this to be the case and awarded Mr. Gore, along with the United Nations Intergovernmental Panel on Climate Change, the 2007 Nobel Peace Prize.

While skeptics remain, the idea that humans are already facing serious challenges because of our reliance on fossil fuel is becoming generally accepted. Changes in global weather conditions as well as scientific reports of rapid glacial and arctic ice melt are evidence that the problem needs serious attention. In 2009, newly elected president Barack Obama promised to take the threat seriously and invest heavily in carbon-free energy. In June of that year, the U.S. House of Representatives passed the American Clean Energy and Security Act, which provides for a mandatory cap-and-trade system for greenhouse gas emissions, as well

BUCKMINSTERFULLERENE

Prior to 1985, chemistry students asked to describe the allotropic forms of carbon most likely would have responded by describing graphite and diamonds. Graphite is probably the form of carbon that is familiar to most people, often in rather large pieces such as lumps of coal. It consists of parallel planes of carbon atoms in which the planes are held together by weak, intermolecular forces of attraction. In the planes themselves, the arrangement of carbon atoms closely resembles a honeycomb, with carbon atoms arranged in interconnecting hexagons. Diamonds are also familiar to most people, although large samples are usually found only in museums. Diamonds have a distinctly different arrangement of atoms than is found in graphite. In diamonds, each carbon atom is bonded covalently with equal strength to four other carbon atoms that are arranged tetrahedrally around it. Because of the equal strengths of the bonds in every direction, diamond possesses great hardness; in fact, it is the hardest mineral known. Graphite and diamonds have the greatest industrial uses of the various allotropic forms of carbon. The other forms that were known prior to 1985 have fewer applications.

It was a great surprise, therefore, when chemists Robert Curl, Jr., and Richard Smalley of Rice University in Houston,

as other measures to ensure a clean energy economy. As of this writing, the bill is awaiting a vote in the Senate. An international climate summit held in Copenhagen in December 2009 failed to produce any binding legislation that would alleviate the problem but served as an indication that governments around the world realize the need to come together to find solutions.

TECHNOLOGY AND CURRENT USES

The uses of carbon are all-encompassing, and synthetic plastics, fibers, and elastomers are extremely important to consumers today. The

Texas, along with Sir Harold Kroto of the University of Sussex in the United Kingdom reported the results of an experiment in which they vaporized carbon with a laser beam and then allowed the carbon clusters that formed to condense. One cluster that Curl, Smalley, and Kroto found contained 60 carbon atoms fitting together in a perfectly symmetrical truncated icosahedron. In fact, the atoms in C_{60} were bonded together in arrangements of 20 hexagons and 12 pentagons, just like the stitching on a soccer ball. *Geodesic* domes also match this pattern. Because the geodesic dome was designed by R. Buckminster Fuller, it was decided to name C_{60} "buckminsterfullerene" in Fuller's honor. Since then, C_{60} and similar clusters have been called "fullerenes," "buckeyballs," "socceranes," and "footballanes" (the last one because what the English call "football" is what Americans call "soccer").

Curl, Smalley, and Kroto shared equally the 1996 Nobel Prize in chemistry for their discovery. Much research has continued since the discovery of fullerenes as chemists, physicists, and materials scientists try to find commercially useful applications for them. One possibility is that C_{60}—because of its sphericity—could have lubricating properties superior to Teflon®, a synthetic polymer used to lubricate hardware that includes chains, cables, moving parts, rusted bolts, wheels, winches, and tools.

accompanying list is representative, but it is not exhaustive of examples of synthetic plastics:

Name of Plastic	Examples of Use
polyethylene	grocery bags
polypropylene	food containers, kitchenware
polystyrene	styrofoam insulation, plastic cups
polyvinyl chloride	credit cards, bottles, plastic pipes
Teflon®	nonstick surfaces on pots and pans
Lucite®	shatter-resistant windows

Bakelite®	radios, telephones, clocks, billiard balls

Synthetic plastics are important in medicine and biotechnology, especially in the form of *prosthetics*—artificial body parts.

The following is a list of examples of synthetic fibers:

Name of Fiber	Examples of Use
nylon	stockings, carpets, fishing line
polyester	fabrics
rayon	fabrics
acrylic	yarn, rugs
Saran™	food wrap

The main example of an elastomer is synthetic rubber. Adding sulfur to synthetic rubber (a process called *vulcanization*) greatly strengthens rubber. Vulcanized rubber is used today to manufacture all tires.

Carbon-based deposits fuel the world's economy. Coal, oil, and natural gas have been the dominant energy sources for more than a century and continue to be so. Fossil fuels are nonrenewable resources, but wood and charcoal are still important fuels, especially in developing countries. Diamonds consist of mostly pure carbon and are treasured worldwide in jewelry and in industry. Other forms of mostly pure carbon include *coke,* which is used in steelmaking, and *carbon black,* which makes a black color in applications such as coloring rubber tires. In addition to Teflon, graphite is a dry lubricant, with simple applications such as lubricating door locks. Carefully fabricated graphene—a single layer of graphite—may soon replace silicon in improved electronic circuits. Due to its unique structure, graphene is the strongest substance known.

The pharmaceutical industry relies on products like cholesterol-lowering drugs, blood-thinners, and flu and pneumonia vaccines—just a few of the huge number of brand name and generic drugs on the market—all of which are carbon-based compounds.

Many archaeological artifacts are dated using carbon-dating methods. The ratio of carbon 14 to carbon 12 in wooden beams, tool and weapon handles, textiles, and charcoal from cooking fires can be used to estimate the ages of the artifacts.

3

Nitrogen: In the Atmosphere

In the form of nitrogen molecules (N_2), nitrogen is the most abundant element in Earth's atmosphere. Comprising 79 percent of air by volume, nitrogen is essential for plant growth and for the formation of protein and nuclear acid molecules. The nitrogen molecule itself, however, is relatively chemically inert, which is why there is such an abundance of it. Microorganisms living in *symbiotic* relationships with the roots of plants have to *fix* the nitrogen. That is, these *nitrogen-fixing bacteria* have to convert N_2 into forms that can be assimilated by plants—forms like the NO_3^-, NO_2^-, and NH_4^+ ions. In turn, animals fulfill their requirement for nitrogen by eating plants.

Nitrogen exists in a variety of compounds in the atmosphere, including nitrous oxide (or "laughing gas," N_2O), nitric oxide (NO),

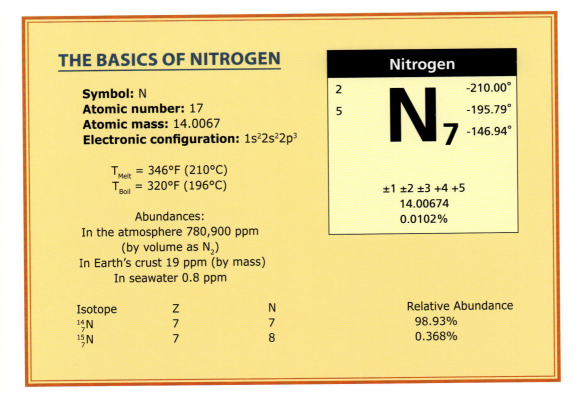

THE BASICS OF NITROGEN

Symbol: N
Atomic number: 17
Atomic mass: 14.0067
Electronic configuration: $1s^2 2s^2 2p^3$

T_{Melt} = 346°F (210°C)
T_{Boil} = 320°F (196°C)

Abundances:
In the atmosphere 780,900 ppm
(by volume as N_2)
In Earth's crust 19 ppm (by mass)
In seawater 0.8 ppm

Nitrogen			
2			-210.00°
5	N		-195.79°
	7		-146.94°
	±1 ±2 ±3 +4 +5		
	14.00674		
	0.0102%		

Isotope	Z	N	Relative Abundance
$^{14}_{7}N$	7	7	98.93%
$^{15}_{7}N$	7	8	0.368%

nitrogen dioxide (NO_2), dinitrogen trioxide (N_2O_3), dinitrogen tetroxide (N_2O_4), and dinitrogen pentoxide (N_2O_5). In the atmosphere, none of these compounds are beneficial to living organisms. N_2O is a greenhouse gas and is harmful to stratospheric ozone. NO_2 is a toxic air pollutant; in addition, it is responsible for harmful ozone formation in the troposphere. N_2O_5 dissolves in water vapor to form nitric acid (HNO_3), one of the harmful components of *acid rain*.

The topics that will be covered in this chapter include the origin and discovery of nitrogen; the nitrogen cycle; nitrogen's role in fertilizers, explosives, and air pollution; and technological uses of nitrogen.

THE ASTROPHYSICS OF NITROGEN

Nitrogen, the ubiquitous gas that comprises 78 percent of Earth's atmosphere, was, like all elements, initially forged in stars. Until recently, it was believed that all nitrogen formed in intermediate- to high-mass stars—those that have a large proportion of heavy elements—via a process called the carbon-nitrogen-oxygen (CNO) cycle. These must be fairly young stars, as the heavy elements would have been produced by supernova explosions, which did not happen in the early universe.

As described in chapter 2, carbon at the core of a star is produced by fusion of helium nuclei. Once carbon is formed, the CNO process can take place in the following steps:

$$^{12}_{6}C + ^{1}_{1}H \rightarrow ^{13}_{7}N + energy$$

$$^{13}_{7}N \rightarrow ^{13}_{6}C + e^{+} + ^{1}_{6}n$$

$$^{13}_{6}C + ^{1}_{1}H \rightarrow ^{14}_{7}N + energy$$

$$^{14}_{7}N + ^{1}_{1}H \rightarrow ^{15}_{8}O + energy$$

$$^{15}_{8}O \rightarrow ^{15}_{7}N + e^{+} + ^{1}_{0}n$$

$$^{15}_{7}N + ^{1}_{1}H \rightarrow ^{12}_{6}C + ^{4}_{2}He + energy$$

Until recently it was not understood how the carbon could meet up with the hydrogen in order to make the nitrogen.

In older, heavier stars, the carbon core begins to collapse once the fusing helium nuclei in the core become so sparse that distance keeps them from interacting. The collapsing core again provides heat that spreads outward to the helium shell, allowing fusion from helium to carbon to occur there. The helium shell now contains carbon that can come in direct contact with the outermost shell, which is hydrogen. So this is how the carbon meets the hydrogen in massive stars. But there is another way, only recently recognized. Nitrogen has been observed in the oldest stars, those that were the first to form after the big bang. These stars were too lightweight for the collapse temperature to get high enough to fuse carbon. So what happened then? How could nitrogen have come into being?

The answer is that rotation must be taken into account. Low-mass stars are by nature more compact than heavier ones, and the difference has a strong effect on *angular velocity* variation; there are different velocities of rotation, depending on the distance from the center of the star. This angular velocity gradient allows for extensive mixing of gases between different layers, somewhat like what happens when soup is stirred. Some of the carbon from the core is carried outward and meets up with the hydrogen in the outer shell, where the CNO process can begin far from the core.

DISCOVERY AND NAMING OF NITROGEN

The ancient Greeks believed there were only four elements—earth, air, fire, and water. Consequently, air was considered a single gas, and centuries passed before people began to realize that air is, in fact, a mixture of several gases. The story of the discovery of air's components begins with the theory of *phlogiston.*

Georg Ernst Stahl (1660–1734), born in Anspach, Bavaria, studied medicine and chemistry and became a professor of medicine at the University of Halle. He is remembered today mostly for his work in the field of chemistry.

In Stahl's time, combustion and corrosion phenomena were poorly understood. Stahl introduced the concept of *phlogiston,* an intangible substance that he believed to be contained in various common substances such as charcoal and metals. The term *phlogiston* comes from the Greek word *phlogos,* which means "flame." According to Stahl, charcoal is rich in phlogiston. When charcoal burns, it loses its phlogiston to the atmosphere. Similarly, when metals rust (become what is called their *calx*), the metals lose their phlogiston to the atmosphere. When the calx is reduced back to the metal again, the calx reabsorbs phlogiston from the atmosphere. An important point here is that Stahl felt that the atmosphere normally does not contain phlogiston; instead, the atmosphere is just a carrier of phlogiston.

What should have been an obvious flaw with the phlogiston theory is that charcoal and other fuels lose weight when they burn, but metals gain weight when they rust or become a calx. If the theory of phlogiston were correct, these observations would imply that there must be two kinds of phlogiston—one with positive weight and one with negative weight! Despite this contradiction, the theory of phlogiston dominated chemistry for almost the next 100 years.

In 1756, Scottish chemist Joseph Black (1728–99) described a gas produced by heating limestone ($CaCO_3$). Black observed that this gas did not support combustion. He called the gas *fixed air,* but did not try to identify it. He also observed that when a candle burned in a closed container, the candle eventually was extinguished (as we now know, because the flame depleted all of the oxygen in the container). Black's fixed air was now in the container. When Black removed the fixed air, another gas remained that also would not support combustion. Black could not explain these results.

Years later, Black turned his experiments over to his student Daniel Rutherford (1749–1819). Both men believed in the theory of phlogiston and therefore tried to explain their results in terms of phlogiston. In 1772, Rutherford called the gas that remained after combustion was complete—and after the fixed air had been removed—"phlogisticated air." Today, phlogisticated air is known as nitrogen, and historians give Rutherford the credit for discovering it. Joseph Black's "fixed air" is known today as carbon dioxide.

The theory of phlogiston was finally abandoned after the discovery of oxygen. Nevertheless, phlogiston played an important role in the attempt to explain the chemistry of common oxidation and reduction processes.

THE NITROGEN CYCLE: HOW PLANTS BREATHE

Nitrogen is the only element that occurs more abundantly in the atmosphere than in the oceans, freshwater systems, soil, and rocks combined. Nitrogen molecules are so chemically *inert* that they tend not to react with anything else in the atmosphere. What makes nitrogen so passive?

The answer lies in the chemical bonding between the nitrogen atoms. In the nitrogen molecule, the atoms are held together by a triple bond, symbolized by writing $N \equiv N$. A triple bond is an extremely strong bond. Oxygen molecules are held together by a double bond, symbolized $O=O$, and hydrogen molecules are held together by only a single

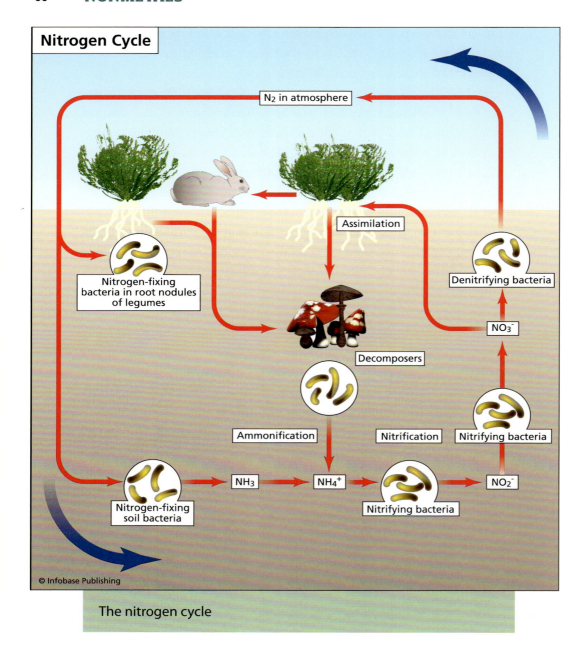

Nitrogen Cycle

N₂ in atmosphere

Assimilation

Nitrogen-fixing bacteria in root nodules of legumes

Denitrifying bacteria

NO₃⁻

Decomposers

Ammonification   Nitrification   Nitrifying bacteria

Nitrogen-fixing soil bacteria   NH₃   NH₄⁺   Nitrifying bacteria   NO₂⁻

© Infobase Publishing

The nitrogen cycle

bond, symbolized H–H. Separating atoms in a nitrogen model requires almost twice the energy that is needed to separate atoms in an oxygen molecule, and slightly more than twice the energy needed to separate atoms in a hydrogen model. Until the atoms in a molecule are broken apart from each other, those atoms are not available to be incorporated into other molecules. The energy to break bonds in molecules usually comes from energetic collisions between molecules. Since the energy generated in collisions usually is insufficient to break the triple bond in a nitrogen molecule, the molecules tend to stay intact. For example, N_2 and O_2 molecules are colliding with each other constantly, yet very few collisions result in the formation of NO or N_2O. In the atmosphere, the most likely way to dissociate nitrogen molecules is during lightning storms, when bolts of lightning supply the energy that is required to break the triple bonds.

Despite the difficulty of making nitrogen atoms available to form other compounds, the fact remains that nitrogen is an essential element required by all living organisms. Photosynthesis cannot take

In the atmosphere, the most likely way to dissociate nitrogen molecules is during lightning storms when bolts of lightning supply the energy that is required to break the triple bonds. (*NOAA's National Severe Storms Laboratory [NSSL] Collection*)

place in the absence of the appropriate nitrogen-containing nutrients. All amino acids (of which protein molecules are made) and all nucleic acids (DNA, RNA) require atoms of nitrogen in their molecular structure. (In fact, nitrogen constitutes 5 percent of the weight in a human being.) Since neither plants nor animals can incorporate nitrogen atoms directly from atmospheric nitrogen gas, whence do plants and animals get their nitrogen?

Because plants and animals cannot assimilate N_2 directly, atmospheric nitrogen must first be *fixed,* that is, it must be converted into chemical compounds that plants can assimilate. (Animals do not assimilate any nitrogen compounds directly; all of the nitrogen that animals use comes from the foods animals eat.) In the natural environment, nitrogen fixation is accomplished by nitrogen-fixing bacteria that are either free living or that exist in a symbiotic relationship in the root nodules of certain plants like legumes (peas, beans, clover, alfalfa, etc.). In the first step—occurring both in soils and in water—nitrogen-fixing bacteria called *azotobacters* ("azo" referring to nitrogen) convert atmospheric nitrogen into ammonia (NH_3). The ammonia then dissolves in water to form ammonium (NH_4^+) and hydroxide (OH^-) ions. These are the first steps in what scientists call the *biogeochemical cycle* of nitrogen.

Nitrification by other bacteria ("nitrobacters") converts NH_4^+ into nitrite (NO_2^-) and nitrate (NO_3^-) ions, which are assimilated by plants through their root systems. The nitrogen atoms become part of the proteins the plants manufacture and that animals obtain by eating plants. Extra nitrogen becomes a waste product that is excreted in urine as the compound urea. When plants and animals die and undergo decomposition, nitrogen compounds are converted back into NH_3, some of which returns to the atmosphere, but much of which is reassimilated by other organisms. In fact, decomposition of organic wastes and reassimilation of nitrogen by other plants is the principal source of nitrogen in most ecosystems. In the atmosphere, NH_3 reacts with O_2 to revert to N_2. Denitrifying bacteria can also convert nitrate ions back into nitrogen molecules, which return to the atmosphere.

Plants must have nitrogen to grow. In the beginnings of agriculture, animal manure was the primary source of nitrogen. Much later in history, fertilizer was introduced. The main component of fertilizer

A sodium nitrate mining ghost town in northern Chile in January 2000, where only the curator remains (*©Julio Etchart/The Image Works*)

is nitrogen, usually in the form of ammonia, an ammonium ion salt, a nitrate ion salt, or ammonium nitrate (NH_4NO_3), which incorporates both ions. Originally, these compounds were obtained in one of two ways: (1) by destruction of hoofs, horns, and other inedible animal products; or (2) by distillation of coal, which contains nitrogen because of coal's plant origin. To some extent, nitrogen fertilizer has also been derived from *guano,* which is derived mostly from the dung of sea birds and bats.

The development of industrial processes to manufacture ammonia synthetically has been a tremendous boon to the fertilizer industry. At the beginning of the 20th century, there was concern about whether fertilizer production could keep pace with the problem of growing enough food to feed the world's increasing population. The first large-scale process was developed in 1913 by Fritz Haber (1868–1934) and Carl Bosch (1874–1940), two German chemists. What is referred to as the *Haber-Bosch process* involves manufacturing ammonia directly from its elements: nitrogen and hydrogen. Typically, the reaction combines

N_2 and H_2 at a temperature of 930°F (500°C) and a pressure of around 200 atmospheres in the presence of an iron catalyst. Under these conditions, the yield of ammonia is about 98 percent. After the ammonia has formed, the mixture of gases is cooled to a temperature where ammonia liquefies—about −28°F (−33°C)—separating the ammonia from the remaining gaseous nitrogen and hydrogen. The nitrogen and hydrogen can be recycled to produce more ammonia. More than ca. 11 million tons (10 million tonnes) of ammonia are manufactured in the United States annually and more than ca. 5.07 million tons (4.6 million tonnes) in Canada.

Unfortunately, the Haber-Bosch process is very energy intensive. In fact, it has been estimated that perhaps 1 percent of the world's total energy consumption comes from the production of ammonia by the Haber-Bosch process. Since Haber and Bosch first developed their process, scientists have tried to find less energy-intensive methods to make ammonia. The problem that needs to be resolved is that it is very difficult to break the triple bond holding the nitrogen atoms in N_2 together. In 2007, researchers reported a breakthrough: A catalyst consisting of single tantalum atoms attached to a silica surface is capable of splitting N_2 and combining the nitrogen atoms with hydrogen atoms to make ammonia molecules.

THE NO$_X$ PROBLEM

Normally, atmospheric nitrogen (N_2) is relatively chemically inert because of the strong triple bond holding the nitrogen atoms together. However, combustion processes—especially those occurring in internal combustion engines—take place at high enough temperatures that some nitrogen gas can react with oxygen gas to produce small amounts of nitric oxide (NO) by the following chemical reaction:

$$N_2 \text{ (g)} + O_2 \text{ (g)} \rightarrow 2 \text{ NO (g)}$$

Nitric oxide is a toxic, colorless gas. In a motor vehicle engine, only a small amount of nitric oxide is formed. There are, however, very large numbers of motor vehicles (cars, trucks, buses, SUVs, motorcycles) on the road, so that collectively the emissions of nitric oxide can be quite large.

Photochemical smog has serious health effects; for example, it can trigger respiratory ailments like asthma and emphysema. *(iofoto/Shutterstock)*

Nitric oxide reacts with atmospheric oxygen to produce nitrogen dioxide (NO_2) by the following reaction: $2 \, NO \, (g) + O_2 \, (g) \rightarrow 2 \, NO_2 \, (g)$. Nitrogen dioxide is a yellowish brown gas that is very irritating to the eyes and lungs. In addition, nitrous oxide (N_2O, also known as "laughing gas") is a product of both fires and microbial activity. Because these compounds—NO, NO_2, and N_2O (and others)—contain nitrogen and oxygen in various ratios, collectively they are referred to as "NO_x" (pronounced "Nox"). Nitrous oxide is relatively inert in the atmosphere and tends not to be involved in the chemistry of the lower atmosphere (the *troposphere*).

The oxides of nitrogen—principally NO and NO_2—are referred to as "primary" air pollutants because they are chemicals that result directly from combustion processes. Subsequently, nitrogen dioxide reacts with sunlight to *dissociate* into nitric oxide and oxygen atoms by the following reaction: NO_2 (g) + sunlight → NO (g) + O (g). Single oxygen atoms in the atmosphere are extremely chemically reactive. They combine easily with diatomic oxygen molecules (O_2) to form ozone (O_3) by the following reaction: O (g) + O_2 (g) → O_3 (g). Ozone is an example of a "secondary" air pollutant because it is not produced directly in engines, but is formed in the atmosphere from primary air pollutants.

Because the molecules that compose fuels are not completely burned to carbon dioxide and water, fires and motor vehicles are sources of carbon monoxide. Emissions also contain fragments of *volatile organic compounds* (VOCs), which in the atmosphere become peroxides, aldehydes, and other poisonous compounds. Together, ozone, NO_x, carbon monoxide, and these organic pollutants comprise what is called *photochemical smog* because the production of so many of the chemicals is driven by the interaction of sunlight with NO_2. Photochemical smog has serious health effects; for example, it can trigger respiratory ailments like asthma and emphysema. Toxic chemicals and ground-level ozone also damage trees, crops, and some synthetic materials like nylon and rubber.

Photochemical smog first became an issue in the Los Angeles area; therefore, photochemical smog is often referred to as "Los Angeles–type smog." Another type of smog is called "industrial smog," or "London smog" (because it was first observed in London, England, during the reign of King Henry VIII). The word *smog* is a contraction of the terms *smoke* and *fog* because the first London smog quite literally was due to a combination of smoke and fog. London smog is more typical of stationary sources that burn coal. Because sulfur is an impurity in coal, London smog may also include sulfur dioxide (SO_2), a particularly noxious air pollutant, along with carbon monoxide. Recent severe episodes of London smog occurred in October 1948, in Donora, Pennsylvania, in which 20 people died and 6,000 people became ill, and in 1952, in London, England, when 3,000 people died in what experts called a "Killer Fog." Sometimes, to distinguish the two types of air pollution, rather than referring to them both as "smog," the Los Angeles type is called photochemical "haze" rather than "smog."

The emission of nitric oxide also leads to the formation of nitric acid (HNO_3), through a series of complex chemical reactions in the atmosphere driven by the energy in sunlight. The emission of sulfuric dioxide leads to the formation of sulfuric acid (H_2SO_4), through similar chemical reactions. Nitric acid and sulfuric acid are both what chemists call *strong acids.* Prevailing winds may carry these acids downwind for hundreds of miles before they finally return to the surface of Earth in the form of *acid deposition.* Acid deposition may be in the form of acid rain, snow, sleet, dust, or other particulate matter. In areas where sedimentary rocks (e.g., limestone) and deep topsoils are found, the rocks and soil are capable of *buffering,* or *neutralizing,* the extra acid. However, in areas characterized mostly by igneous rocks and thin top soils, that buffering capacity is absent. In those areas, the acid can release metals that are otherwise bound in the soils and raise the acidity of aquatic ecosystems, with deleterious effects on fish, amphibians, insect larvae, and aquatic plants. In the United States and Canada, regions that are most susceptible to the effects of acid deposition are upstate New York, New England, and Ontario.

In 1970, the United States Congress created the Environmental Protection Agency (EPA). In the same year, the Clean Air Act was passed by Congress and signed by President Richard Nixon. Under the authority granted to the EPA by the Clean Air Act, the EPA has established emission standards for the most common air pollutants: particulate matter, ground-level ozone, carbon monoxide, the oxides of sulfur, the oxides of nitrogen, and lead (Pb). The individual states may establish their own, stricter emission standards, but they cannot establish standards that are laxer than the standards set by the federal government. For example, in California, air quality is monitored by the California Air Resources Board (CARB); California's air quality standards are stricter than the national standards. These emission standards apply to industrial sources, electrical power plants, and motor vehicles, and even to smaller pollution sources such as gasoline stations, paint shops, and dry cleaners.

To meet the emissions standards for motor vehicles, cleaner fuels were introduced. In addition, the catalytic converter was invented. Catalytic converters contain a metal catalyst—usually platinum, rho-

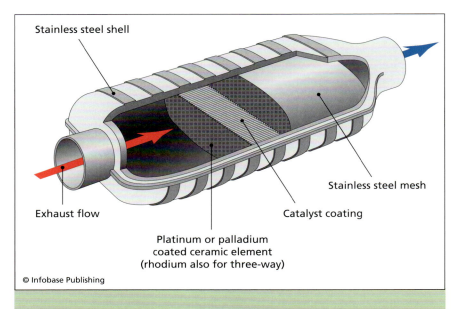

Stainless steel shell

Stainless steel mesh

Exhaust flow

Catalyst coating

Platinum or palladium
coated ceramic element
(rhodium also for three-way)

© Infobase Publishing

A pellet-type catalytic converter. Catalytic converters contain a metal catalyst—usually platinum, rhodium, or palladium—that removes CO, NO_x, and VOCs from motor vehicle exhaust by converting them to CO_2, H_2O, and N_2. (Source: www.quickhonda. net/exhaust.htm)

dium, or palladium—that removes CO, NO_x, and VOCs from motor vehicle exhaust by converting them to CO_2, H_2O, and N_2. Catalytic converters have been extremely successful; emissions of air pollutants are 90 percent lower today than they were in 1970. (A side benefit is that the reduced emissions of CO have made accidental—or intentional—carbon monoxide poisoning a rare event.)

Beginning in the 1920s, the compound tetraethyl lead was added to gasoline to improve performance. Unfortunately, the lead showed up in motor vehicle exhaust, where its inhalation posed a serious health hazard. In response to this problem, the Clean Air Act mandated a reduction in lead emissions. It was also discovered that lead in motor vehicle exhaust tended to coat the surface of the catalyst in the catalytic converter, rendering the catalyst ineffective. To remedy both problems, unleaded gasoline was introduced in 1974. Leaded gasoline continued to be sold, but was eventually phased out. Since that time, the levels of lead in the atmosphere have dropped significantly.

NITROGEN NARCOSIS AND DECOMPRESSION SICKNESS

Nitrogen narcosis refers to the narcotic effect of excess nitrogen dissolved in brain tissue. The problem is very familiar to scuba divers because they rely on compressed air, which, like our atmosphere, is 78 percent nitrogen. Increased pressure as divers descend forces a disproportionate fraction of nitrogen into the body's system. Neurons in the brain are particularly affected, leading to impaired thinking, which is noticeable to most divers by the time they reach 100 feet (30.5 m) of depth; the effects range from a mild feeling of intoxication to unconsciousness. Unlike alcohol intoxication, however, nitrogen narcosis can set in very quickly and is alleviated as soon as pressure decreases sufficiently, with no hangover effects.

Decompression sickness refers to other physiological effects that can occur during ascent to the water's surface—the decompression phase. If the decompression rate is too fast, the absorbed nitrogen will preferentially form bubbles. This process has variable results, depending on the size and location of the bubbles in the body. If they develop in the muscles or spine, extreme pain and muscle contraction can result—a consequence that has earned the term "the bends." Bubbles can also block the bloodstream or cause pressure in the brain.

Divers attempt to avoid bubble growth by making periodic decompression stops while rising, thereby allowing the nitrogen to return into solution in the blood. A relatively new procedure called the Reduced Gradient Bubble Model, developed by Bruce Wienke, a master diver and physicist at Los Alamos National Laboratory, can inhibit bubbles by restricting their access to combinations of gases that tend to enhance growth. Divers can surface in a shorter period of time by using differently mixed gases and even pure oxygen, depending on whether the decompression stop is deeper or shallower.

If a diver does surface too quickly, some hospitals have *hyperbaric chambers,* where the sea depth pressures are reapplied and slowly reduced while the victim breathes oxygen as pure as 100 percent.

N₂ USE IN AUTOMOBILE TIRES

Racing cars and aircraft use pure nitrogen in their tires instead of the normal air admixture consisting of approximately 78 percent nitrogen, 21 percent oxygen, and 1 percent argon. There is almost always a small percentage of water vapor in the mix, however, depending on the ambient humidity when the air is compressed. Pure nitrogen is drier and filtered of particulate matter. The water vapor in the normal air admixture is very good at absorbing and emitting heat, which causes tires to slightly contract and expand. This can affect high-performance cars because pressure changes of a fraction of a pound can change the handling capabilities on high-speed turns. Large aircraft, too, have a brief period of time on takeoff and landing when handling is crucial.

Permeability of the various gases also must be considered. Oxygen molecules, with a dynamic diameter of 0.346 nm, are minutely smaller than nitrogen atoms at 0.364 nm. Water molecules are even smaller, with an effective diameter of 0.32 nm. This means that oxygen gas and water vapor are slightly more permeable through rubber, and so can escape from a tire more easily. In addition, oxygen is corrosive to most materials and can degrade the inner liner of the tire.

Each of these effects alone is small, but taken together can contribute over time to tire deflation and lowered gas mileage. Pure nitrogen can alleviate that problem, but so can regular checks and maintenance of tire pressures.

Unlike nitrogen, which is effectively inert in the body, oxygen is used up quickly in tissues and cells, so bubbles do not have time to form.

Astronauts, too, have to worry about decompression illness when leaving a spacecraft or when the craft loses pressure. Typical spacesuits are designed to maintain about 30 percent atmospheric pressure (inflating suits to higher pressures makes them unwieldy). To prepare for the

Liquid nitrogen is used in laboratories and hospitals to keep machines, drugs, and tissues at a super-cold temperature.
(Charles D. Winters/Photo Researchers, Inc.)

lowered pressure, astronauts need to spend several hours breathing 100 percent oxygen to flush nitrogen from body fluids.

TECHNOLOGY AND CURRENT USES

Vaporized liquid nitrogen easily and rapidly absorbs heat energy and is used as a coolant for preservation of food, blood, and tissue samples. It is also convenient and relatively safe for use in scientific research relating to the behavior of materials at low temperatures.

Because pure N_2 gas is dry, easily produced, and safe to breathe, it is the gas of choice for automobile airbags and nonfluorescent lightbulbs. Because it is less sensitive to pressure changes than air, pure nitrogen gas is typically used to fill race car and aircraft tires.

Nitrogen also has many important uses as a chemical component. Nitrogen controls many pharmaceutical reactions, for example. It also allows photosynthesis to take place. The main component of fertilizer is nitrogen, usually in the form of ammonia; an ammonium ion salt; a nitrate ion salt; or ammonium nitrate (NH_4NO_3), which incorporates both ions.

While nitrogen is generally inert, when a nitrogen molecule forms by the fusion of two nitrogen atoms, a good deal of energy is given off—360 joules per mole. Therefore, explosives that incorporate nitrogen are very effective, and bombs can be made from fertilizer.

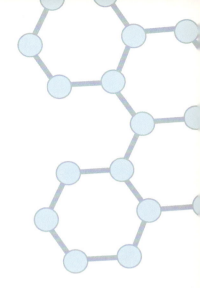

4

Phosphorus: Fertilizers, Photosynthesis, and Strong Bones

Phosphorus is element number 15 and the 11th most abundant element in Earth's crust. Usually found in the form of phosphates, it exists in two allotropic forms: white phosphorus and red phosphorus. Because white phosphorus is poisonous, it is too hazardous for common use. A large amount of red phosphorus, in the form of P_4S_3, is combined with lead dioxide (PbO_2) to coat the tips of matches. When struck, the lead dioxide rapidly oxidizes the phosphorus and the match bursts into flames. With safety matches, the phosphorus and an abrasive coat the side of the box; the matches will not ignite unless they are struck on the box in which they came.

An essential component of DNA, phosphorus is one of the three most important elements found in fertilizers (along with nitrogen and

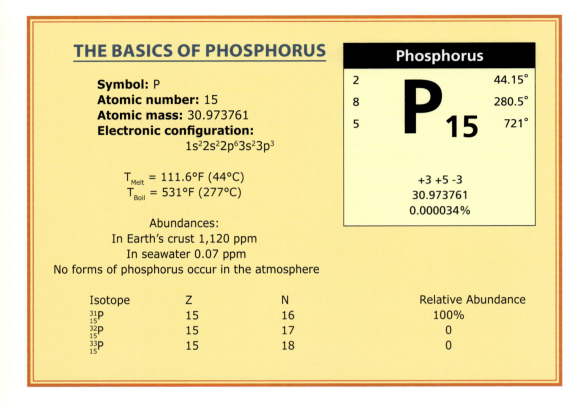

THE BASICS OF PHOSPHORUS

Phosphorus

Symbol: P
Atomic number: 15
Atomic mass: 30.973761
Electronic configuration:
$1s^2 2s^2 2p^6 3s^2 3p^3$

T_{Melt} = 111.6°F (44°C)
T_{Boil} = 531°F (277°C)

Abundances:
In Earth's crust 1,120 ppm
In seawater 0.07 ppm
No forms of phosphorus occur in the atmosphere

2
8
5

P 15

44.15°
280.5°
721°

+3 +5 -3
30.973761
0.000034%

Isotope	Z	N	Relative Abundance
$^{31}_{15}$P	15	16	100%
$^{32}_{15}$P	15	17	0
$^{33}_{15}$P	15	18	0

potassium). In addition to DNA, phosphorus is present in the molecule *adenosine triphosphate* (ATP), which is a source of energy in the body. As an important nutrient, phosphorus is a major component of bones and teeth.

THE ASTROPHYSICS OF PHOSPHORUS

Phosphorus is scarce in the universe and in stars, though most exhibit at least weak phosphorus spectral lines. In the outer atmosphere of a typical star like the Sun, the ratio of phosphorus to hydrogen is less than 3×10^{-7}. Yet this seemingly unimportant element is absolutely key to the formation of life on Earth and is much more abundant in biological systems, where the ratio of phosphorus to carbon is as high as 0.02. DNA and RNA depend on it.

How did this element so crucial to life form in stars, and how did it become available in such abundance for the evolution of organisms on Earth? Again, the process of fusion in stellar cores is the recipe, but

Elemental phosphorus comes in red and white forms.
(Richard Treptow/Photo Researchers, Inc.)

phosphorus needs a new twist. Phosphorus 31 can be produced in any star at least 15 times as massive as the Sun. The high mass is necessary to produce the extreme temperatures needed at the core for fusion of such a heavy element. About 10 percent of all stars meet this criterion.

In such stars, phosphorus is an end result of the fusion of two oxygen atoms

$$^{16}_{8}O + {}^{16}_{8}O \rightarrow {}^{31}_{15}P + \text{proton}$$

This process occurs in heavy stars after neon-burning in the core. But it does not give the answer to the abundance of phosphorus on Earth or the other planets. Although oxygen burning does produce phosphorus in heavy stellar cores, further nucleosynthesis uses up almost all of it to produce, in the end, a star composed of 90 percent silicon and sulfur. So another mechanism must be responsible for phosphorus production and distribution throughout the universe.

Supernova explosions seem to be the answer. The shock of the explosion allows even outer shells of the star's atmosphere to heat up to temperatures in the range of 2 billion K—hot enough to allow neon to fuse. This explosive neon burning and its subsequent reactions result in the production of ^{31}P in the star's atmosphere. The matter created in this way is ejected with enough force to overcome the star's diminished gravitational attraction, and the elements are free to sail into the universe, to be collected and incorporated in newly forming stars and planets.

Still, the density of phosphorus in the interstellar medium is extremely low, and chemists and life scientists strive to understand how it was present on early Earth in high enough concentrations for life to begin and evolve. Iron meteorites, corroded at the surface to form a phosphate salt layer, may have rained down on the planet with sufficient frequency to provide the needed surplus, particularly in suitable freshwater environments. Even so, *prebiotic* RNA synthesis from phosphates would have required chemical conditions that seem unlikely to have been available. According to some scientists, circumstances could have been more conducive to formation of glyoxylic acid ($C_2H_2O_3$), whose ionized form, *glyoxylate,* may have been the prebiotic forerunner of RNA. This is an exciting area of research that continues to be debated and to grow.

DISCOVERY AND NAMING OF PHOSPHORUS

The phenomenon of *phosphorescence* was known in ancient times. Elemental phosphorus, however, was not isolated and recognized as such until the 17th century.

Hennig Brand (1630–1710), born in the 17th century in Hamburg, Germany, is often regarded as the last of the alchemists. Little information is available about his personal life, including the exact dates of his birth and death. (The spelling of his first and last names varies with the source: Some spell his first name "Henning," while others spell it "Hennig," and some spell his last name "Brandt" and others as "Brand.") What *is* known is that Brand was a physician and chemist whose wife's name was Margaretha. They lived in Hamburg, Germany, and his discovery of phosphorus most likely occurred in 1669. Phosphorus was only the fourth element—after arsenic, bismuth, and zinc—to be added to the list of elements known to the ancient world. Brand's discovery brought him a considerable amount of fame.

Brand's method of discovering phosphorus was unique; he found it by distilling his urine. (Since phosphorus is found in DNA, teeth, and bones, it is natural that some phosphorus would be excreted as a waste product.) Brand did not leave a record of exactly what it was he was looking for in his urine, but what he found was a white, waxy substance that glowed in the dark.

THE *P* IN PHOTOSYNTHESIS

Scientists debate the methods by which life began on Earth, but there is no question that, without phosphorus, plants and animals could not exist. All plants rely on phosphorus as a building block to produce glucose, the food that fuels growth of leaves, flowers, fruits, and seeds via the process known as photosynthesis.

Photosynthesis takes place in cell structures called *chloroplasts*. These cells contain chlorophyll, the pigment that allows sunlight to be trapped so its energy can be used for growth. Chlorophyll is most efficient at absorbing light of blue and red wavelengths, but not the green or yellow parts of the visible spectrum, which is why most plants appear green or greenish-yellow.

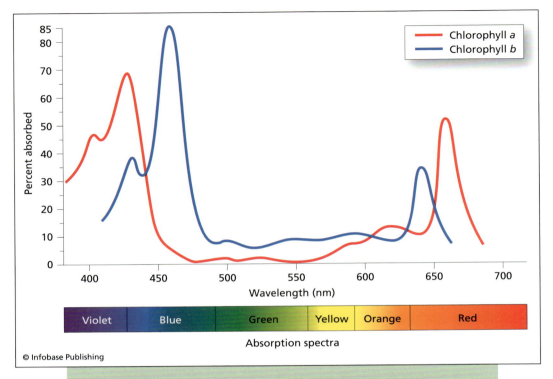

The absorption spectra of chlorophyll A and chlorophyll B

© Infobase Publishing

There are two distinct stages by which photosynthesis proceeds: the light-dependent phase and the light-independent (formerly called "dark") phase.

The light-dependent phase requires sunlight to initiate energy transfer. Most plants can transform light into electrical and then chemical energy with an impressive 95 percent efficiency. Sunlight promotes the electrons in chlorophyll from lower- to higher-energy quantum states of the molecule. Only recently have these excitations been understood to occur under *resonant* conditions that allow coherent coupling among electronic states (similar to two-electron excitations in the negative hydrogen ion; see chapter 1). The coherent nature of the excitations allows for nearly instantaneous transfer of energy to neighboring electrons, which significantly speeds the subsequent construction of energy-storing molecules.

The energy-storing molecules, bearing the rather daunting names "reduced nicotinamide adenine dinucleotide phosphate" (NADPH) and "adenosine triphosphate" (ATP), are needed for use in the second or light-independent stage of photosynthesis. Considerable energy is stored in the bonds of ATP. For example, when ATP is converted into adenosine phosphate (ADP) with the release of one phosphate group (as shown in the following equation), about 30 kJ of energy are released per mole of ATP molecules.

$$ATP + H_2O \rightarrow ADP + phosphate + energy$$

In the presence of ambient CO_2, NADPH provides hydrogen molecules, while ATP provides the energy to make glucose ($C_6H_{12}O_6$). Glucose chains are used for storing energy, as they are too bulky to move through plant membranes. When the plant needs energy, it must convert glucose to sucrose, a process that is also aided by ATP. Sucrose molecules transport sugar from the leaves to the nonphotosynthetic parts of a plant.

Oxygen molecules are the other major by-product of the light-independent stage of photosynthesis. Plants provide the oxygen needed for animal respiration on Earth (see chapter 5).

HIGHER YIELDS: PHOSPHORUS AND AGRICULTURE

The three most important plant nutrients are nitrogen (N), phosphorus (P), and potassium (K). Any element required by plants can be the *limiting factor* that determines the success of plant growth, but nitrogen and phosphorus are the two elements most likely to be limiting. Whereas the primary source of nitrogen is ultimately the atmosphere, the primary natural source of phosphorus is rocks that contain the phosphate ion (PO_4^{3-}), especially apatite (calcium phosphate). In addition to the direct importance of phosphorus to plants, phosphorus is also indirectly important in that the ability of legumes to fix nitrogen depends on the availability of phosphorus.

Although phosphorus tends to be relatively abundant in soils, most of it is found in insoluble minerals such as aluminum and iron phosphate and is inaccessible to the plants. In agriculture, therefore, about 10 million tons (9.07 million tonnes) of phosphoric acid are produced in the United States every year, about 80 percent of which is used in the manufacture of fertilizers. Even then, much of the phosphorus that is applied to crops quickly becomes locked into the soils and cannot be taken up by the plants, so that a quantity of phosphorus in excess—sometimes a very large excess—of what plants actually require has to be applied.

There are a variety of phosphate ions that include $H_2PO_4^-$, HPO_4^{2-}, and PO_4^{3-}. The relative abundance of these ions depends on the *pH* of the soil solution. (The pH of a solution measures the relative acidity of the solution. Low pHs are acidic; high pHs are basic. A neutral solution has a pH of 7.)

In general, plants more readily assimilate the $H_2PO_4^-$ ion. The concentration of $H_2PO_4^-$ is highest at slightly acidic pHs, which suggests

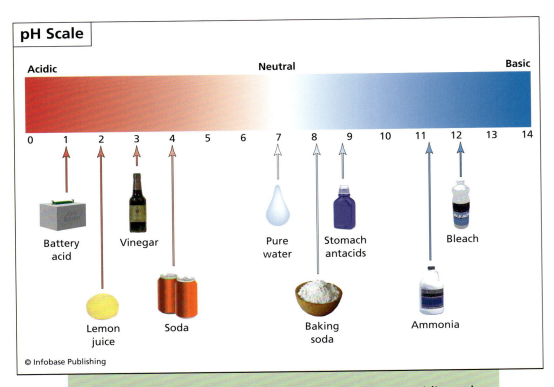

pH Scale

Acidic   Neutral   Basic

0   1   2   3   4   5   6   7   8   9   10   11   12   13   14

Battery acid   Vinegar   Pure water   Stomach antacids   Bleach

Lemon juice   Soda   Baking soda   Ammonia

© Infobase Publishing

The pH Scale. A low pH means that a solution is more acidic, and a high pH means that a solution is more basic; a pH of 7 means the solution is neutral.

that phosphorus should be more available to plants at acidic pHs. Other factors, including the presence of mineral cations (e.g., Ca^{2+}, Fe^{2+}, Fe^{3+}, Mn^{2+}, and Al^{3+}), also affect the ability of plants to assimilate phosphorus, however, because these cations tend to precipitate phosphate, rendering phosphorus unavailable to plants. Thus, the exact conditions that best favor phosphorus uptake are fairly complicated.

Commercial fertilizers typically express the amounts of nitrogen, phosphorus, and potassium they contain as N-P-K percentages. Homeowners tend to want a lush, green lawn, which means a greater need for nitrogen. Therefore, a lawn fertilizer might give its N-P-K content as 22-2-14, which means, by weight, the fertilizer is 22 percent N, 2 percent P, and 14 percent K. A specification of 31-3-8 would be 31 percent N, 3 percent P, and 8 percent K. On the other hand, rose gardeners are less interested in growth of their rose plants, but are more likely to desire healthy flower production and strong root systems, which means a need for less nitrogen and more phosphorus. Therefore, a rose fertilizer might have an N-P-K content of 10-10-10, which means, by weight, that it contains 10 percent N, 10 percent P, and 10 percent K.

Although adding excessive amounts of phosphates to the environment could result in excessive, unwanted plant and algal growth in aquatic environments, so much of the excess amount of phosphate in fertilizers is locked into the soil that relatively little phosphate leaches into ponds and streams adjacent to farmlands. Excess phosphate from detergents is potentially a much greater problem for the environment.

INSECTICIDES

In general, insecticides can be grouped according to whether they are persistent (i.e., long lasting) or nonpersistent (i.e., they break down quickly in the environment into relatively harmless compounds). DDT (dichlorodiphenyltrichloroethane) is probably the best-known persistent insecticide. Even though it is actually less potent than many other insecticides on the market, DDT lasts for years in the environment and becomes more concentrated in the tissues of animals as it works its way up the food chain.

Among the earliest insecticides were naturally occurring poisons like arsenic, strychnine, thallium, copper sulfate, and nicotine. The effects of these compounds tend to be relatively short-lived. One of the first groups of nonpersistent synthetic insecticides included the organic phosphate compounds, derivatives of phosphoric acid. *Organophosphate* compounds can cause acute poisoning and, over time, adversely affect an animal's nervous system enough to lead to paralysis.

Nerve impulses are transmitted by chemicals like acetylcholine. Once an impulse has been transmitted, an enzyme called cholinesterase destroys the acetylcholine molecule so that it does not keep sending more impulses. Organophosphate compounds prevent cholinesterase from destroying acetylcholine. Because impulses are being continuously transmitted, severe spasms occur that can result in death.

Examples of organophosphate compounds include the following: diazinon, parathion, and malathion. Diazinon is used to eradicate insect pests on farms where crops like lettuce, almonds, broccoli, and spinach are grown, and it is used in home gardens—all this despite its toxicity to humans. The adverse effects of diazinon include the following: It is a carcinogen, a neurotoxin, and an acute toxin; it damages reproductive organs; and it inhibits fetal development. In residential use, any poisonous substance poses a threat to small children, pets, and persons with allergic susceptibilities to chemicals or weak immune systems. In agricultural use, poisonous substances pose a threat to farm workers. Excess insecticide can appear in drinking water. In addition, fruits and vegetables sold in supermarkets could be contaminated with harmful residues. Because of diazinon's adverse health effects, beginning in 2005, the EPA began banning all residential uses of diazinon. In 2007, the EPA also began to impose restrictions on agricultural applications. In September 2009, the agency ruled diazinon to be unsafe for home use and placed new restrictions on retailers (typically home-and-garden stores): All diazinon-containing products must be removed from retail shelves within three years and supplies may not be replenished.

Parathion can accidentally be inhaled, swallowed, or absorbed through the skin. It is too toxic to children and pets for any residential applications. Symptoms include dizziness, muscle twitching and weakness, nausea, vomiting, and abdominal cramping. High exposure can cause respiratory failure and death. Parathion, however, still has agricultural applications, where only outdoor spraying occurs and where its residue on fruits and vegetables can decompose before the produce goes to market.

Malathion is applied to numerous fruit, vegetable, and berry crops. Homeowners can purchase malathion to apply to grass, shrubs, flowers, and trees. It has widespread use in the control of boll weevils (which are particularly damaging to cotton crops), Mediterranean fruit flies, and mosquitoes. Malathion has even been used to treat head lice. Target pests also include such diverse species as ants, scorpions, aphids, caterpillars, cockroaches, earwigs, crickets, fleas, millipedes, ticks, and silverfish. Of the 15 million pounds of malathion applied in the United States annually, two-thirds is used just to eradicate boll weevils.

PHOSPHORESCENCE WITHOUT PHOSPHORUS

The ability of a material to absorb light energy, store it, and emit it later at a different wavelength is called phosphorescence. After taking in light, if a typical molecule of a material de-excites immediately back to the ground state by re-emitting light, this is called fluorescence. But a molecule in a phosphorescent material can transition into what is called a *metastable* state, relaxing into the state by vibration rather than by giving off radiation. So some of the absorbed energy is used, but not given off as light. This comparatively stable molecular configuration comes about because of changes to electron orbit and spin orientation. The molecule cannot stay in the state forever, and gradually the electrons jump from the metastable state back to the ground state, giving off light in the process. This is called phosphorescence. Both fluorescence and phosphorescence are methods of photo-luminescence, as photons both induce the process and are produced by it.

Zinc sulfide paints have long been used for phosphorescent emergency signage, but are not very bright and may dim after only a few minutes. The newer pigments containing alkaline earth aluminates can be brighter by as much as a factor of 10 and are now used almost exclusively for signs. Researchers continue to experiment with new phosphorescent molecules that could be used in novel light-emitting devices.

Though the name comes from first observations (ca. 1670 C.E.) of glowing phosphorus, phosphorus itself is not phosphorescent. That long-observed effect, explained by R. J. Van Zee and A. M. Khan in 1974, instead comes from oxygen reacting with phosphorus at the surface of the material to form the light-emitting molecules HPO or P_2O_2. Photons

Glow sticks glow because oxygen reacts with phosphorus at the surface of the material to form light-emitting molecules.
(Joel Sartore/National Geographic/Getty Images)

are not needed to initialize the reaction. This process is, therefore, correctly termed *chemical luminescence.*

PHOSPHATES AND THE ENVIRONMENT

Most natural water supplies contain at least some calcium ions (Ca^{2+}). Calcium ions tend to precipitate with the chemicals in soaps and detergents, giving the familiar soap scum and bathtub "rings" in bathrooms and kitchens. A particular problem in the case of laundry detergents is the deposits left on clothing. Phosphates, in the form of *polyphosphates,* were added early in the history of laundry detergents because calcium would then precipitate with phosphate instead of with the detergent. When phosphate was used, clothes were cleaner and fresher. Unfortunately, because phosphate is also a primary plant nutrient, the enrichment by phosphate in wastewater resulted in greater *productivity,* i.e., a greatly increased growth of aquatic plants. The proliferation of aquatic plants, along with *algal blooms,* tended to choke waterways. In addition, as the plants subsequently died, the decomposers present in the ecosystems consumed such large quantities of dissolved oxygen that other aquatic organisms—insects, fish, amphibians—were left with insufficient oxygen and died as a result. Decaying vegetation also produces foul odors. Added nitrogen has the same effect; the two nutrients together can wreak havoc on a lake or pond.

This cycle of adding nutrients to an aquatic ecosystem, the increased productivity, and the subsequent decomposition of plant materials and oxygen depletion is called *eutrophication*. Eutrophication can also be a natural process, although in an unpolluted environment the result is more likely to be limited to increased algae growth, usually blue-green algae.

By the 1960s, local communities had begun to ban the use of detergents that contained phosphates. Finally, the problem with phosphates in detergents became so acute that the big three detergent makers—Procter & Gamble, Lever Brothers, and Colgate-Palmolive—removed most phosphates from detergents.

This pond is in an advanced stage of eutrophication, thereby making it an unfriendly habitat for many native species. *(Dr. Terry McTigue, NOAA, NOS, ORR)*

TECHNOLOGY AND CURRENT USES

In the United States, the fertilizer industry produces large quantities of phosphorus-containing products. During the first decade of the 21st century, manufacturers yearly produced about 9 million tons (10 million tonnes) of diammonium phosphate [$(NH_4)_2HPO_4$], 3.6 million tons (3.3 million tonnes) of ammonium sulfate [$(NH_4)_2SO_4$], and 11 million tons (10 million tonnes of phosphoric acid (as P_2O_5, which, when mixed with water, forms H_3PO_4). Besides supplying phosphorus for plant growth, the first two products are also important sources of nitrogen.

The hardness of water is due to calcium and magnesium ions (Ca^{2+} and Mg^{2+}), which combine with soap to form insoluble precipitates. Phosphates can *sequester,* or tie up, calcium and magnesium, thereby acting to *soften* the water. Trisodium phosphate (TSP, or Na_3PO_4) is used as a cleanser in synthetic detergents. Painters often use TSP to wash previously painted walls in preparation for new coatings of paint. To inhibit rusting, automobile surfaces are sometimes treated with phosphoric acid (H_3PO_4) before being painted. Dicalcium phosphate [$Ca_3(PO_4)_2$] is a common polishing agent in toothpaste.

Bakers use compounds that contain phosphate as leavening agents in cake mixes, self-rising flour, and baking powder.

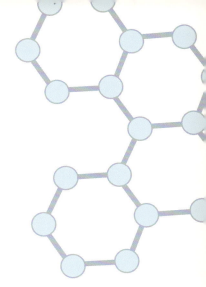

5

Oxygen: From Flames to Pollution

Oxygen is element number 8 and the most abundant element in Earth's crust. The pure element exists in two allotropic forms: O_2 and O_3 (ozone), which are both gaseous species under average atmospheric conditions. Diatomic oxygen gas, O_2, composes 21 percent of the atmosphere and is an essential biological element. Ozone composes a thin layer of the stratosphere and is essential to protecting life on the surface of Earth from otherwise harmful ultraviolet radiation. Unfortunately, in the troposphere, ozone is an air pollutant.

Oxygen is found in a number of compounds that are essential to life. Water is made of oxygen and hydrogen. Carbohydrates, amino acids, nucleic acids, and—to a small extent—lipids all contain oxygen. Just as oxygen is necessary to support combustion, oxygen is necessary

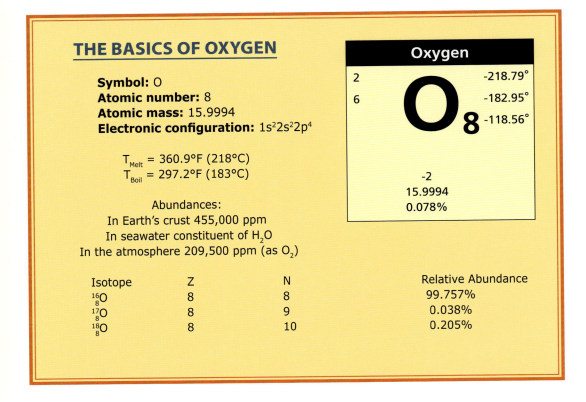

THE BASICS OF OXYGEN

Symbol: O
Atomic number: 8
Atomic mass: 15.9994
Electronic configuration: $1s^2 2s^2 2p^4$

T_{Melt} = 360.9°F (218°C)
T_{Boil} = 297.2°F (183°C)

Abundances:
In Earth's crust 455,000 ppm
In seawater constituent of H_2O
In the atmosphere 209,500 ppm (as O_2)

Isotope	Z	N	Relative Abundance
$^{16}_{8}O$	8	8	99.757%
$^{17}_{8}O$	8	9	0.038%
$^{18}_{8}O$	8	10	0.205%

Oxygen

2
6

O_8

-218.79°
-182.95°
-118.56°

-2
15.9994
0.078%

for cellular respiration, which is a slow, controlled form of combustion. Without oxygen, life as we know it would not exist.

OXYGEN IN STARS AND ON EARTH

In stars more than about twice the mass of the Sun, the CNO cycle (see chapter 3) is responsible for the synthesis of oxygen in the stellar core, via the fourth step in the cycle:

$$^{14}_{7}N + ^{1}_{1}H \rightarrow ^{15}_{8}O + \text{energy}.$$

This isotope of oxygen is fairly unstable, however, and soon spontaneously decays in the following manner:

$$^{15}_{8}O \rightarrow ^{15}_{7}N + e^+ + ^{1}_{0}n.$$

So while this method of oxygen production is prolific and fairly direct, it cannot sustain an oxygen population that will be ejected into space.

A less efficient, but more reliable source of interstellar oxygen originates in the red-giant phase of stars, which occurs after the useable hydrogen has fused to helium in the core (see chapter 2). In this stage, the so-called triple-alpha process produces ^{12}C, but a side effect is that the carbon can interact with the available helium nuclei, so that

$$^{12}_{6}C + {}^{4}_{2}He \rightarrow {}^{16}_{8}O + energy.$$

Oxygen 16 is a stable isotope and survives because the subsequent possible interaction

$$^{16}_{8}O + {}^{4}_{2}He \rightarrow {}^{20}_{10}Ne + energy$$

is suppressed due to a nuclear spin effect (an explanation of which is beyond the scope of this text).

In stars of mass greater than about eight times the mass of the Sun, neon-burning temperatures can be reached, and the neon produced can interact with photons (energy) to form oxygen via the reverse reaction of that shown above

$$^{20}_{10}Ne + energy \rightarrow {}^{16}_{8}O + {}^{4}_{2}He.$$

The formation of $^{16}_{8}O$ via either of the above reactions also occurs in the stellar atmospheres of expanding supernovae, which allows the explosive distribution of oxygen into the cosmos.

Throughout interstellar space, there are on average 320 oxygen atoms for every million hydrogen atoms—about 0.032 percent O/H. Yet Earth's atmosphere is 21 percent oxygen. How did this surplus arise?

As the planet's formation was in its very early stages, before a solid core had formed, the atmosphere was probably very similar to the interstellar medium (ISM)—mostly H_2 and He gas. After the core hardened and volcanoes formed, eruptions spewed a combination of gases, including H_2 and H_2O, CO and CO_2, S_2 and SO_2, Ne, Cl_2, NH_3 (ammonia), and CH_4 (methane), but not O_2. A small fraction, however, originated from the breakup of atmospheric water molecules by ultraviolet light.

It is important that oxygen gas was extremely scarce in Earth's early atmosphere. If it had been more abundant, life as we know it probably could not have evolved. Oxygen interferes with the chemistry that allows formation of amino acids—the building blocks of proteins—and of the most primitive bacteria. It was a fine balance, however, as a 1 to 2 percent oxygen content is needed to form an ozone layer of a thickness adequate to protect newly forming life from solar ultraviolet radiation. If there was not enough ozone, life most likely formed first underwater, as water attenuates UV radiation up to 99 percent within the first half meter of the surface. (See chapter 6 for a discussion of how pyrite samples or Yellowstone bacteria may answer the question of oxygen content in the early atmosphere.)

Some current research, however, finds that the molecules forming the basis of life tend to need a cycle of wet and dry phases, which would suggest a shoreline origin. Others studies indicate that a more friendly biosynthetic environment would be between layers of rock, such as mica, where important nutrients would have been more readily available.

However life began, it is the source of our current abundance of oxygen in the atmosphere, as free oxygen is a by-product of photosynthetic organisms (see chapter 4). Early cyanobacteria (formerly called blue-green algae), which developed 2½ to 4 billion years ago during the Archean age, were the first photosynthesizers.

DISCOVERY AND NAMING OF OXYGEN

Ancient Greeks believed that air is a single substance and therefore an element. That belief dominated *alchemy* for centuries. Fire was also believed to be an element; a true understanding of the nature of fire eluded people until the end of the 18th century. Questions to be answered included the following: Why do some objects burn and others do not? What is the nature of combustion? The answers to these questions required the discovery of oxygen, which ultimately would prove to be one of the most important discoveries in the entire history of chemistry and indeed would mark the transition from alchemy to the era of modern chemistry.

Under the Greek notion of the four essential elements, a combustible object must contain within itself some essence of fire. When that object burns, it releases that essence. For many years, that essence was believed to be sulfur, but in the 17th century sulfur was replaced by a hypothetical new substance.

In the latter part of the 17th century, the German physician and chemist Georg Ernst Stahl (1660–1734) proposed that the essence of combustion is a substance that he called *phlogiston,* from the Greek meaning "to set on fire." Stahl used the concept of phlogiston to explain two seemingly dissimilar phenomena: combustion and the rusting (*calcination*) of metals. In the case of combustion, inflammable objects contain phlogiston, which is released to the atmosphere when the object burns. Since phlogiston should have weight, this would explain why wood, coal, or any other fuel loses weight when it burns. The combustion of substances like sulfur and phosphorus would be explained in the same manner. Metals also would contain phlogiston. According to Stahl's theory, when a metal rusts, the metal gives off phlogiston to the atmosphere. What remains is the metal's calx. When the calx itself is heated, phlogiston would be reabsorbed from the atmosphere and the calx is converted back into the original metal. In either case Stahl felt that air itself was not involved in these processes. Instead, air was just a temporary reservoir for phlogiston.

A problem with this theory is that it contradicts itself. If a burning fuel or a rusting metal both give off phlogiston to the atmosphere, they should both lose weight in the process. When a metal calx is converted back into the metal itself, phlogiston is gained, suggesting that a pure metal should weigh more than its calx. Experiments showed that combustible materials like wood, coal, sulfur, and phosphorus, in fact, do lose weight as they burn. However, the opposite happens with metals. Metals *gain* weight as they become their calx, and a calx *loses* weight as it reverts to the pure metal. Although the experimental results were incompatible with the theory, chemists at that time were not interested in the quantitative aspects of their science. Thus, Stahl and the next several generations of chemists adopted the theory of phlogiston and tried to explain chemical reactions in terms of it. Resolution of this dilemma required the discovery of oxygen and its role in both the combustion of fuels and the rusting of metals.

Credit for the first isolation of oxygen gas itself is usually given to the Swedish chemist Carl Wilhelm Scheele (1742–86), who, in 1771, prepared oxygen by heating mercuric oxide (HgO), which upon heating decomposes into liquid mercury and gaseous oxygen. Scheele, however, did not publish his results until after similar results were reported by Joseph Priestley in 1777.

Joseph Priestley (1733–1804) was a Unitarian minister who served a church in Leeds, England, from 1767 to 1773. Priestley was just as interested in science as he was in theology; among other experiments, he studied the properties of gases. Unaware of Scheele's prior preparation of oxygen gas, in 1774, Priestley prepared oxygen by the same method—first heating mercury so that the mercury would combine with oxygen in the air, and then heating the mercuric oxide that was formed to produce mercury and pure oxygen, which he collected and used in his studies to understand its role in supporting combustion. Priestley also conducted several experiments that consisted of placing mice in closed containers both with and without plants. In the absence of plants, the mice died rather quickly (because they ran out of oxygen). In the presence of plants, the mice thrived because the plants continued to produce oxygen by photosynthesis. Thus, oxygen's role in respiration was established.

It was not Priestley, however, who gave oxygen its name. Still believing in the theory of phlogiston, Priestley called his new gas dephlogisticated air. Priestley reported his discovery to Antoine Lavoisier (1743–94). Lavoisier repeated Priestley's experiments and recognized that the chemistry of this new gas explained the discrepancies in the phlogiston theory. In 1783, Lavoisier ushered in the era of modern chemistry with the publication of his oxygen theory of combustion, stating that Priestley's dephlogisticated air was in fact a fundamental component of air and the element responsible for the combustion of fuels and the rusting of metals. Ironically, Lavoisier mistakenly believed that acids had to contain this same element. Consequently, Lavoisier named Priestley's gas *oxygen* from the Greek meaning "acid-former." Lavoisier also demonstrated that air is roughly one-fifth oxygen and four-fifths nitrogen.

Priestley—who by this time had moved to Birmingham, England— was not a particularly popular clergyman. In 1791, his liberal theology and sympathy with both the American and the French Revolutions resulted in the burning down of both his church and his home. In 1794, Priestley immigrated to the United States and spent the last 10 years

British chemist Joseph Priestley isolated and examined many new gases, including hydrogen chloride, nitrous oxide, oxygen, and nitrogen. *(Sheila Terry / Photo Researchers, Inc.)*

of his life in Northumberland, Pennsylvania, continuing to write both theological and scientific works. His home in Northumberland is now listed in the National Register of Historic Places and is open to the public for tours.

THE CHEMISTRY OF OXYGEN: FROM ANTIOXIDANTS TO FREE RADICALS

A human can live for a month without food, a week without water, but only 15 minutes without oxygen. Oxygen is essential for life processes to occur. Fortunately, oxygen is all around us; it is one-fifth of the air we breathe and an important element in many of the rocks and minerals in Earth's crust. All combustion reactions require oxygen. Combustion, however, is not just the burning of wood and fossil fuels, but also describes cellular respiration, the essential biochemical process by which all living organisms—plant or animal—derive the energy necessary to grow, move, and reproduce.

Oxygen is a very versatile element chemically. Oxygen readily forms chemical bonds to all of the other elements in the periodic table except the halogens, the noble gases, and a few of the most nonreactive metals.

The source of oxygen gas in the atmosphere is photosynthesis, a biochemical process by which green plants ("green" because they contain the compound chlorophyll) take in carbon dioxide, water, and the energy in sunlight and manufacture sugar. Oxygen gas is a by-product of that process, but a vital by-product, for without oxygen gas, the vast majority of the animals that live on Earth would not exist. Animals—and plants—assimilate oxygen gas and use it to burn sugars to provide energy.

As was discussed in the section on the chemistry of carbon, oxygen is an important element in biologically important compounds including carbohydrates, proteins, and nucleic acids. These molecules all contain oxygen atoms with hydrogen atoms bonded to them (as well as nitrogen atoms that have hydrogen atoms bonded to them in the case of proteins and nucleic acids). As a result, these classes of biological compounds can form both *intramolecular* hydrogen bonds (bonds between different parts of the same molecule, e.g., as occurs in DNA) and *intermolecular* hydrogen bonds (bonds, for example, between a sugar molecule and a water molecule). Consequently, sugars and amino acids are miscible with water; in other words, they tend to form aqueous solutions quite easily. Fats and oils contain almost no oxygen atoms (being almost entirely just carbon and hydrogen). Therefore, fats and oils are nonpolar and *immiscible* with water, hence the saying, "Oil and water do not mix."

The combination of oxygen with other elements is an example of a process chemists call oxidation. In the case of combustion reactions, we say that the fuel is being oxidized and that the oxidation process is something beneficial. In the case of corrosion reactions, metals are being oxidized, as in the case where iron (or steel) combines with oxygen to form "rust" (ferric oxide, Fe_2O_3), which is an undesirable process. Most metals are obtained from deposits of their ores where the metal—for example, copper, iron, zinc, aluminum, or lead—occurs in combination with elements such as oxygen, sulfur, or a halogen, or with a combinations of elements as in a silicate, an arsenate, or a carbonate.

To extract the pure metal from its ore, the metal must be *reduced* from its combined state to its free state. Reduction is the opposite of oxidation: Combining with oxygen (or a similar element) is an example of oxidation; removing oxygen (or a similar element) is an example of reduction. Reducing metals from their combined states in ores to their free states as neutral elements involves very energy-intensive processes and therefore is very expensive. Recycling used materials is desirable because recycling bypasses the reduction step, saving both energy and money.

Oxidation and reduction play important roles in human physiology. Oftentimes fragments of molecules called *free radicals* are present in our tissues and bloodstream. Free radicals tend to be unstable and thus are very chemically reactive because they try to change into substances that are more stable. Therefore, free radicals react chemically with other substances in our bodies, causing oxidative stress in the form of damage to our cells, protein molecules, and genetic material (DNA). Free radicals originate from the metabolic products of the foods we eat, from inflammation of tissues, and from environmental sources such as pollution, X-rays, cigarette smoke, and alcohol. To counter the effects of free radicals, *antioxidants* are found in the foods we eat—and in nutritional supplements we may choose to add to our diets—in the form of enzymes (for example, catalase and peroxidase) and vitamins (commonly beta-carotene and vitamins C and E). Other antioxidants include coenzyme-Q10 and substances called *phytochemicals* that occur in plants.

Free radicals also occur in the natural environment and are responsible for many of the chemical reactions that take place in soils, in aquatic systems, and in the atmosphere. For example, many of the oxidation reactions occurring in the atmosphere are due to single oxygen atoms (O), hydroxyl radicals (HO), and peroxy radicals (HO_2)—which are formed by the action of sunlight on water and other molecules in the atmosphere—and oxides of nitrogen (NO or NO_2), which are formed during combustion processes.

As was discussed in the section on the chemistry of hydrogen, the important compound water (H_2O) contains hydrogen and oxygen. One molecule of water contains two hydrogen atoms and one oxygen atom. Because the shape of a water molecule is "bent"—and oxygen is the second most *electronegative* element in the periodic table—water is an extremely *polar* molecule. Water is so polar that water molecules are strongly attracted to each other by the intermolecular force of attraction of hydrogen bonding. The presence of hydrogen bonding in water— and hydrogen bonding that can occur between water molecules and other molecules in contact with water—gives water many of its unique properties.

All told, oxygen is absolutely essential to life as we know it. Without oxygen there would also be no water. Without O_2 or H_2O, Earth would be a cold, hard, lifeless rock orbiting the Sun, not the planet teeming with life whose picture from outer space is so beautiful.

OZONE ABOVE AND BELOW

Ozone is the name for the O_3 molecule, which exists in two distinctly different locales in our planetary system. One is far above the Earth in the stratosphere, and one is at the surface in the troposphere. One affects animals and plants beneficially; the other affects them unfavorably.

Ozone near the Earth's surface mainly occurs in the smog of cities. It is the major component of pollution, resulting from interactions among anthropogenic and natural emissions of exhaust, fuel vapors, volatile organic compounds, and solvents. Animals and plants take in O_3 through ordinary gas exchange with the atmosphere. Reviled for its deleterious effects on human lungs, ground level ozone causes asthma and other serious lung congestion problems. In 1963, recognizing the

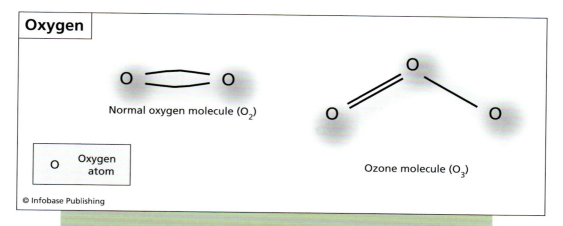

Oxygen

Normal oxygen molecule (O₂)

O   Oxygen atom

Ozone molecule (O₃)

© Infobase Publishing

Diagrams of O_2 and O_3: The common oxygen molecule has a double bond between its oxygen atoms, whereas ozone has one single bond and one double bond.

risks to human health, the Environmental Protection Agency (EPA) set limits on automobile emissions, but it was not until 1971 that a specific standard was set for ozone production compliance at 0.8 parts per million (ppm) averaged over a one-hour period. This was upgraded in 1997 to 0.8 ppm averaged over an eight-hour period, offering greater public protection. While dozens of U.S. cities still experience noncompliance several times per year, concentrations have been decreasing fairly steadily since 1980. In April 2007, the EPA issued new and more restrictive standards proposed as low as 0.7 ppm with an achievement goal of the year 2020.

Though maximally produced in urban pockets, ground-level ozone can also seriously harm vegetation even in rural areas, depending on prevailing winds and weather patterns. High-ozone environments inhibit chlorophyll production, diminishing a plant's capacity to make and store sucrose and starch for its own use. This has a weakening effect, leaving it less resistant to disease and harsh conditions. Death of cells or the entire plant can result. U.S. crop losses run into multiples of the $100-million range annually owing to these factors, with peanuts, soybeans, and cotton being particularly vulnerable. Greenhouses often

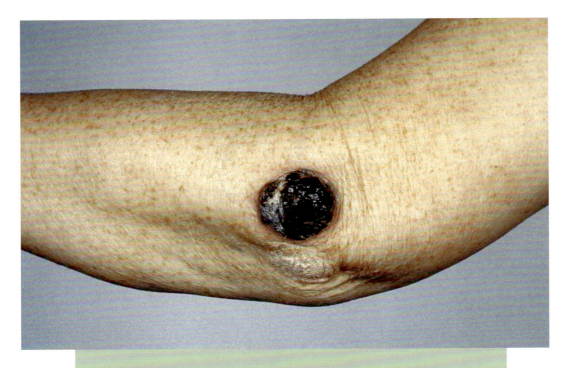

Melanoma, a dark-pigmented, usually malignant tumor, can result from excessive exposure to ultraviolet light. *(Biophoto Associates / Photo Researchers, Inc.)*

use activated charcoal filters to decrease tropospheric ozone delivered to crops. American forests are also at risk: Aspen and black cherry trees have been monitored as being exceptionally sensitive.

Stratospheric ozone, on the other hand, which is present in the atmosphere at altitudes ranging from 3.7 miles (6 km) up to around 4–13 miles (30 km), protects humans, animals, and plants from the Sun's harmful ultraviolet (UV) wavelengths. At these altitudes, O_3 is formed by interaction of cosmic rays with naturally occurring NO_x in the upper atmosphere. Ozone is efficient at absorbing UV radiation and reemitting it in the infrared wavelengths—as heat—which is why the stratosphere's temperature increases with height and why smog-infested cities can become so-called heat islands. Since radiation wavelengths around 300 nanometers have been shown to present a high risk of skin cancer, especially to people whose skin has a low melanin content, the high-level ozone acts as a shield to protect the health of humans as well as virtually all other organisms at the planet's surface.

This natural shield has been damaged, however, by certain halogenated hydrocarbons used popularly as refrigerants and aerosol propellants starting around the 1930s. Effective legislation, in the form of the Montreal Protocol, is alleviating the problem, but the "ozone hole" will not recover completely until far in the future.

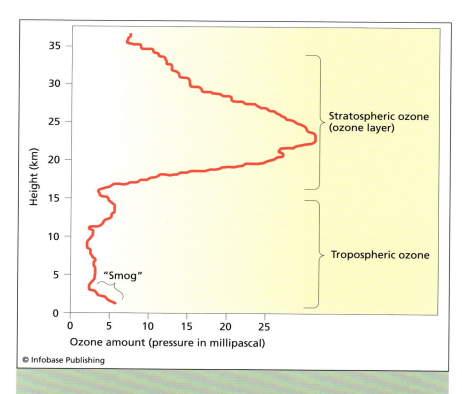

© Infobase Publishing

Ozone as a function of altitude *(Source: www.en.wikipedia.org/ wiki/File: Atmospheric-ozone.svg)*

COMBUSTION, FIRE, AND EXPLOSIONS

Combustion can be strictly defined as a heat-producing (exothermic) reaction that occurs when a fuel and an oxidant are brought together. The simplest combustion reaction is one that is used in many rocket engines.

$$2 H_2 + O_2 \rightarrow 2 H_2O + \text{heat energy}$$

A RECENT TREND: OXYGEN BARS

When people have breathing trouble related to such problems as smoke inhalation, high-altitude *hypoxia,* or asthma, emphysema, and other lung diseases, carefully monitored oxygen supplementation may be necessary. Other serious conditions that need this kind of medical attention include heart attack, *hypoxemia,* gangrene, and bone damage resulting from radiation treatment. A particular need for supplementary oxygen that can require hyperbaric chamber treatment is decompression sickness (see nitrogen narcosis section in chapter 3).

According to the U.S. Food and Drug Administration, oxygen administered in this manner—from one person to another—falls under the category of prescription drug, though the agency allows state guidelines to take precedent, perhaps because a little excess oxygen has not been shown to harm anyone. This has allowed an interesting business practice—that of selling "shots" of air with higher oxygen content. The going rate is currently about a dollar a minute at "oxygen bars," where partial pressures typically range from 40 percent to 95 percent O_2.

Selling points include claims that increased oxygen intake can alleviate hangovers, headaches, memory loss, listlessness, and general malaise—none of which has been scientifically demonstrated. Many medical practitioners doubt the possibility of benefit to the system because, although humans normally breathe air that is 21 percent oxygen, about 14 percent is given up on exhalation, indicating that the blood is oxygen-saturated upon absorption of only one-third of what is available. Increasing the concentration,

The fuel is the hydrogen molecule and the oxidant is the oxygen molecule. Water vapor and heat are the by-products. This process is called "complete" combustion because the end product compound consists only of the combustion reactants.

An average campfire is infinitely more complex and by no means qualifies as complete combustion. Because fires on Earth take place in the atmosphere, which is 78 percent nitrogen, and typically burn

therefore, especially over a short time period, should have little, if any, effect.

Some vendors also include aroma as part of the customer's experience, but this can be a risky practice. Food-grade particulates or oils used to add scent are seldom guaranteed as to purity. Oils are especially hazardous, as inhalation can cause inflammation of the lungs.

A group of attendees tries some of the aromatic oxygen at the Airheads Oxygen Bar at the 2005 Biotechnology Industry Organization's annual trade show in Philadelphia.
(AP Photo/Coke Whitworth)

The light of a fire comes from the incandescence of heated soot particles and from chemical luminescence—excited electrons changing energy levels in the carbon and hydrocarbon molecules. *(unlit/Shutterstock)*

hydrocarbon compounds, such as wood, they produce not only water vapor, but also CO_2 and other hydrocarbon compounds, like carbon monoxide and soot, as well as nitrogen oxides. These gaseous by-products are commonly called "flue gases."

Heat from a fire is produced at the surface of the flame, where the oxygen in the air is available to feed the combustion. The heat comes from the energy that is released upon formation of the bonds in H_2O and CO_2. A typical flame burns with a temperature around 1,000° Celsius and gives off about 10 kilojoules of energy per gram of O_2 consumed.

The light of a fire comes from the incandescence of heated soot particles and from chemical luminescence—excited electrons changing energy levels in the carbon and hydrocarbon molecules.

If the combustion proceeds extremely rapidly, a sudden release of gas along with sound energy can result—an explosion. Fireworks and bombs rely on rapid combustion, as do internal combustion engines.

Surprisingly, though humans have intentionally used fire for at least a million years, combustion is not well understood. Recent experiments in space, where gravity is negligible, have shown that colliding pure O_2 and H_2 results in tiny, dim spheres of flame. Theorists are still trying to understand the physics of this phenomenon.

TECHNOLOGY AND CURRENT USES

For human consumption, hospitals and emergency medical technicians use oxygen in life-support situations, and commercial aircraft carry oxygen supplies in the event of an emergency at high altitudes. Enhanced-oxygen air is sold in oxygen bars (see sidebar), but the benefits are dubious.

Oxygen has an important role to play in fuel synthesis. The carefully controlled addition of oxygen to the hydrocarbons in natural gas results in the formation of acetylene, ethylene, and propylene. Reacting oxygen with a hydrocarbon results in the production of syngas, a mixture of carbon monoxide and hydrogen gases. Syngas is a starting material for the synthesis of fuels such as methanol and octane. Liquid oxygen is the oxidizer in rocket fuels. Purified oxygen is used for welding and cutting metals, as with acetylene torches, and in smelting of ores and refining metals.

Oxygen is an important reactant in industrial chemical syntheses. It is used in the production of the pigments titanium dioxide and carbon black and in the glass manufacturing industry.

Because of their oxidizing ability, O_2 and O_3 are used to kill bacteria in sewage treatment systems.

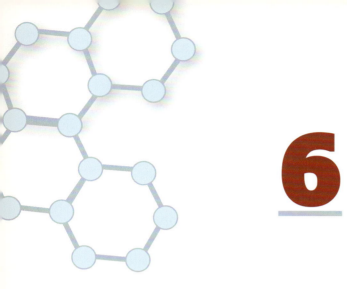

6

Sulfur: In Mythology and Reality

Sulfur is element number 16 and the 16th most abundant element in Earth's crust. Sulfur is one of the few solid elements that can be found in nature in its elemental state as either a yellow solid or flaky material or as a molten liquid. Sulfur's yellow color reminded the alchemists of gold; they attempted repeatedly to convert more common metals to gold by adding sulfur to them.

Several essential amino acids contain sulfur; therefore, sulfur is found in the proteins in human bodies. With the exception of the alkali metals and alkaline earth metals (the first two columns in the periodic table), virtually all of the metals and semimetals form compounds with sulfur, usually as sulfides (containing the ion S^{2-}) or as sulfates (containing the ion SO_4^{2-}). Therefore, because sulfur is such a common element in Earth's crust, there are large deposits of minerals that contain

THE BASICS OF SULFUR

Symbol: S
Atomic number: 16
Atomic mass: 32.066
Electronic configuration:
$1s^2 2s^2 2p^6 3s^2 3p^4$

T_{Melt} = 239°F (115°C)
T_{boil} = 833°F (445°C)

Abundances:
In Earth's crust 340 ppm
In seawater 900 ppm
In the atmosphere < 1 ppm (as SO_2)

Sulphur		
2	**S**	115.21°
8		444.60°
6	$_{16}$	1,041°
	+4 +6 -2	
	32.066	
	0.00168%	

Isotope	Z	N	Relative Abundance
$^{32}_{16}S$	16	16	94.93%
$^{33}_{16}S$	16	17	0.76%
$^{34}_{16}S$	16	18	4.29%
$^{35}_{16}S$	16	19	0
$^{36}_{16}S$	16	20	0.02%

sulfur, including iron pyrite (FeS_2), galena (lead sulfide, PbS), zinc blende (ZnS), and bornite (a mixture of iron and copper sulfide, Cu_3FeS_3). In addition, sulfur occurs in the deposits of sulfate minerals, the most common being gypsum ($CaSO_4 \cdot 2H_2O$).

Most liquids and gases that contain sulfur tend to have rather noxious odors. Notable examples include sulfur dioxide (SO_2), hydrogen sulfide (H_2S), and *mercaptans*—the class of sulfur-based compounds that are responsible for the intense odor of a skunk's spray.

Sulfuric acid (H_2SO_4) is produced in larger quantities than any other industrial chemical. A strong acid that must be handled with care, sulfuric acid is familiar as the electrolyte in the lead storage batteries in motor vehicles.

Red and yellow forms of sulfur are commonly observed in nature.
(*USGS/Bureau of Mines*)

Unfortunately, sulfur is a contaminant of fossil fuels. The combustion of fossil fuels releases noxious oxides of sulfur—sulfur dioxide (SO_2) and sulfur trioxide (SO_3)—into the atmosphere, contributing to respiratory problems, air pollution, and acid rain. Billions of dollars are

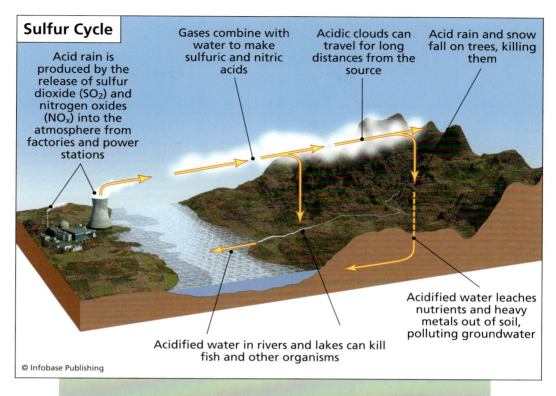

Sulfur Cycle

Acid rain is produced by the release of sulfur dioxide (SO_2) and nitrogen oxides (NO_x) into the atmosphere from factories and power stations

Gases combine with water to make sulfuric and nitric acids

Acidic clouds can travel for long distances from the source

Acid rain and snow fall on trees, killing them

Acidified water leaches nutrients and heavy metals out of soil, polluting groundwater

Acidified water in rivers and lakes can kill fish and other organisms

© Infobase Publishing

Acid rain and the sulfur cycle *(Source: www.filebox.vt.edu/users/ chagedor/biol4684/cycles/seycle.html)*

spent by industry every year to mitigate the harmful effects of sulfur compounds.

SULFUR IN STARS AND ON EARTH

Sulfur is synthesized in a couple of different ways in high-mass stars. The first method is oxygen burning, which occurs after the neon in the center of the star has been used up via neon burning or other processes (see chapter 5), leaving mostly oxygen, magnesium, and α-particles in the core. Oxygen burning requires temperatures of around two billion Kelvin and a density around ten billion kg/m³. Under these conditions, oxygen atoms can fuse, producing sulfur 31 or sulfur 32:

$$^{16}_{8}O + {}^{16}_{8}O \rightarrow {}^{31}_{16}S + {}^{1}_{0}n + \text{energy}$$

$$^{16}_{8}O + ^{16}_{8}O \rightarrow ^{32}_{16}S + \gamma.$$

As another method, oxygen can fuse to make silicon 28, which can fuse with ^{4}He to make the sulfur 32 isotope:

$$^{16}_{8}O + ^{16}_{8}O \rightarrow ^{28}_{14}Si + ^{4}_{2}He + \text{energy}$$

$$^{28}_{14}Si + ^{4}_{2}He \rightarrow ^{32}_{16}S + \gamma.$$

However it is formed, sulfur is distributed throughout the universe via explosive supernovae.

On Earth, sulfur is nearly ubiquitous because of its connection with volcanic activity. Volcanoes and other hot spots, like Yellowstone National Park, emit sulfur-containing gases, notably sulfur dioxide (SO_2). The gases cool and condense, eventually hardening into sulfur crystals and compounds, the purest being lemon yellow in color. (Extraterrestrial sulfur atoms, S_2 and SO_2, have been also detected in the atmosphere of one of Jupiter's moons, Io, which boasts numerous active volcanoes.)

Pyrite (FeS_2) is an extremely common mineral found worldwide that has earned the name "fool's gold" because of its color similarity and reflectivity. In ancient times, the Chinese mined pyrite for its sulfur content, but it is now more economically feasible to recover the sulfur by-product from natural-gas wells. Pyrite is of particular interest to scientists who study the origin of life on Earth because it may provide a clue as to when oxygen became a significant portion of the atmosphere and, therefore, whether early life developed with or without it (see chapter 5). Pyrite deposits are found not only on land, but also on the ocean floor, where they formed from iron particles in seawater bonding with the waste products of sulfate-consuming bacteria. Nowadays, this type of bacteria preferentially ingests sulfates made from the sulfur 32 isotope. If the same was true three billion years ago, then isotopic sulfur ratios from pyrite samples of that era can inform scientists about the oxygen content of the atmosphere, since the stable formation of sulfates is known to require the presence of oxygen. Research continues on this important question.

DISCOVERY AND NAMING OF SULFUR

In regions of high volcanic activity, the element sulfur can be found in pure form. Therefore, no one "discovered" sulfur; ancient people were very familiar with it. The phrase "fire and brimstone" refers to sulfur. Originally, *brimstone* meant the gum of the gopher tree, but later the term *brimstone* was used to refer to anything inflammable, such as sulfur. In fact, medieval alchemists used the word *sulfur* to refer to any substance that could undergo combustion. (An alternate form of spelling *sulfur* is *sulphur.*)

The first reference to brimstone in the Bible occurs in the book of Genesis, chapter 19, verse 24, where the following account is given:

> Then the Lord rained on Sodom and Gomorrah brimstone and fire from the Lord out of heaven.

In Book XXII of Homer's *Odyssey,* when the hero Odysseus returns from the Trojan War, he finds that his house is full of suitors courting his wife. Odysseus slays them, and then calls upon his servant to "bring sulphur . . . that cleanses all pollution and bring me fire, that I may purify the house with sulphur." The Roman historian Pliny described the sulfur mines in Italy and Sicily (regions of the Mediterranean with high volcanic activity). Pliny recorded that there were several practical uses for sulfur, including its use in the following applications: medicine, bleach, matches, and lamp wicks. (Later, when European explorers arrived in the New World, their ability to produce flames with matches was a source of great fascination to the native peoples they encountered, who were still producing fire by friction, such as by rubbing two sticks together or by striking iron-containing rocks with flint.)

Sometimes it is difficult for modern readers to distinguish when earlier writers were using the word sulfur to refer to the element or to refer to any flammable substance. Between the years 300 C.E. and 1100 C.E., almost no chemistry was done in Europe. Instead, the study of chemistry, or *alchemy,* shifted to the Middle East. Probably the most influential Arab alchemist was Jabir ibn Hayyan (ca. 760–ca. 815). Jabir is best known for his studies of the *transmutation* of metals (converting one metal into another metal). Jabir thought mercury was the ideal metal; because mercury is a liquid, it can be purified more completely

In regions of high volcanic activity, the element sulfur can be found in pure form. *(USGS)*

than most metals can. (Mercury can be vaporized easily; the pure vapor can then be collected and condensed into a pure liquid.) Likewise, he thought that sulfur was the perfect example of a combustible substance. Together, he believed that all metals are combinations of mercury and sulfur (or, more precisely, the principles of liquidity and combustibility that mercury and sulfur represented). Gold, for example, seemed to exemplify this idea because gold and sulfur are both yellow in color; therefore, to make gold, one had only to find some material capable of producing the exactly correct mixture of mercury and sulfur. This material that was capable of transmuting one metal into another metal was called an *elixir*. Because it was believed that such a substance would itself have to be dry and earthy, in Europe, this hypothetical transmuting substance was referred to as the *philosopher's stone*.

Of course, no such philosopher's stone was ever discovered. Scientists did in fact eventually learn how to convert one element into another element, but that story is told later in these volumes. In addition, even though the concept of the philosopher's stone was abandoned by the 16th century, the concept that metals contained a combustible principle persisted. This combustible principle was called phlogiston; its story is explained in the sections on the discoveries of nitrogen and oxygen.

THE CHEMISTRY OF SULFUR: KNOWN FOR ITS SMELL

As the aforementioned 16th most abundant element on Earth (approximately 0.1 percent of the crust), sulfur has a rich and varied chemistry. Deposits of free sulfur occur throughout the world, usually associated with regions of volcanism such as Spain, Iceland, Japan, Mexico, Italy, and Sicily. In the United States, sulfur is found primarily in Texas and Louisiana, and to a lesser degree in California, Colorado, Nevada, and Wyoming. (Visitors to the geysers, hot springs, mudpots, and fumaroles at Yellowstone National Park in Wyoming have experienced firsthand the odor of sulfur and its compounds.) In ancient times sulfur was referred to as brimstone.

Sulfur is a component of numerous minerals. Sulfides include iron pyrite ("fool's gold," FeS_2), zinc blende (ZnS), galena (PbS), and cinnabar (HgS). Sulfates include gypsum ($CaSO_4 \cdot 2H_2O$) and barite ($BaSO_4$). These ores are among industry's principal sources of copper, zinc, lead, mercury, calcium, and barium. The sulfates tend to become concentrated in landlocked bodies of salt water. White deposits of sodium, calcium, and magnesium sulfates can be found along the shores of inland lakes like the Great Salt Lake in Utah and Mono Lake in California.

About 90 percent of all sulfur goes into the manufacture of sulfuric acid (H_2SO_4). Sulfuric acid is the world's most abundant industrial chemical, with many applications. About half of all sulfuric acid is converted into compounds used in fertilizers. A portion of the sulfur that does not go into sulfuric acid is used in the vulcanization of rubber tires, a process that strengthens tires and gives them longer lifetimes. In addition, sulfur is used in the manufacture of sulfites, in preserving dried fruit, and for dusting crops to control fungi. About 10 million tons (9.1 million tonnes) of sulfur is recovered annually from deposits in the United States, much of it in the form of hydrogen sulfide gas (H_2S), a contaminant of natural gas, which must be removed from the natural gas. Sulfur is imported from Canada, Mexico, and Venezuela. Canada is a major exporter because Canada is also a major source of natural gas. Between 3 million and 5 million tons (2.7 million and 4.5 million tonnes, respectively) of sulfuric acid are recycled annually from petroleum refining and chemical manufacturing processes.

Sulfur exists in a number of allotropic forms that differ primarily in their crystal structures. The molecular form of sulfur consists of eight sulfur atoms in a ring—denoted S_8. Melting sulfur and then cooling it rapidly creates a plastic form of sulfur. The powdered sulfur often found in school chemistry laboratories is obtained by sublimation and is known as "flowers of sulfur." Sulfur is a nonconductor of electricity and is insoluble in water. The allotropes of sulfur can take on a wide variety of forms, as shown in the table on page 113.

Sulfur is a component of various amino acids such as cysteine and methionine. As such, sulfur becomes part of the protein molecules that make up our bodies and the bodies of other living organisms. The sulfur in the protein of eggs, for example, is responsible for the so-called rotten-egg odor of decaying eggs due to the formation of hydrogen sulfide gas (H_2S). Hydrogen sulfide is a poisonous gas; a concentration of 1 part per thousand in air causes unconsciousness and a level of two parts per thousand is lethal. Hydrogen sulfide in the air is also responsible for

ALLOTROPIC FORMS OF SULFUR

RINGS	ADDITIONAL RINGS
α-S_8 orthorhombic	S_7, S_9, S_{10}, S_{11}, S_{12}, S_{18}, S_{20}
β-S_8 monoclinic	**LONG CHAINS**
γ-S_8 monoclinic	"Plastic" sulfur in several forms
ε-S_6 rhombohedral	

tarnishing sterling silverware. Tarnish consists of a thin layer of silver sulfide (Ag_2S) formed by the reaction between H_2S and Ag. Sulfur in animal tissue is excreted in urine as sulfuric acid.

Several sulfur-containing ions are often familiar from beginning chemistry courses. In one case, two sulfate ions can be linked together to form the peroxydisulfate ion ($S_2O_8^{2-}$), a powerful oxidizing agent. It is often used to oxidize transition metal ions to higher oxidation states, such as the conversion of the manganous ion (Mn^{2+}) to the permanganate ion (MnO_4^-), as shown in the following reaction:

$$8\ H_2O\ (l) + 5\ S_2O_8^{2-}\ (aq) + 2\ Mn^{2+}\ (aq) \rightarrow 10\ SO_4^{2-}\ (aq) + 2\ MnO_4^-\ (aq) + 16\ H^+\ (aq).$$

The solution turns purple as the MnO_4^- forms.

In another set of ions, sulfur replaces an oxygen atom. The prefix "thio" is used to indicate that this has happened. Common examples are the thiocyanate ion (SCN^-), where S has replaced the oxygen in the cyanate ion (OCN^-), and the thiosulfate ion ($S_2O_3^{2-}$), where S has replaced an oxygen in the sulfate ion (SO_4^{2-}). The thiocyanate ion's chemistry is very similar to the chemistry of the halide ions (Cl, Br, and I), since SCN^- combines with the silver ion (Ag^+) to precipitate

AgSCN in analogy to AgCl, AgBr, and AgI. The presence of SCN^- in a solution can be detected by adding a few drops of the ferric ion (Fe^{3+}); the complex ion that results, $FeSCN^{2+}$, is a distinctive blood-red color. This reaction is shown in the following equation:

$$Fe^{3+} (aq) + SCN^- (aq) \rightarrow FeSCN^{2+} (aq).$$

Because fossil fuels are the remains of once-living organisms, fuels like coal contain sulfur. In electrical power plants, sulfur burns along with the carbon in coal, forming sulfur dioxide (SO_2). There also is some sulfur in gasoline, so motor vehicles are also a source of SO_3 (not SO_2 because catalytic converters convert SO_2 into SO_3). Sulfur dioxide and trioxide are both noxious gases (see the section in chapter 3 titled "The NO_x Problem") and two of the main components of so-called London smog. Oxides of sulfur cause respiratory disease and contribute to the formation of haze that reduces atmospheric visibility.

Once it is in the atmosphere, SO_2 may be further oxidized to SO_3 by the following chemical reaction:

$$2\ SO_2 (g) + O_2 (g) \rightarrow 2\ SO_3 (g).$$

SO_3, in turn, is water-soluble and dissolves in water to form H_2SO_4 by the following reactions:

$$SO_3 (g) \rightarrow SO_3 (aq)$$

$$SO_3 (aq) + H_2O \rightarrow H_2SO_4 (aq),$$

where "aq" means "aqueous," (i.e., that the substance is dissolved in water). Together with nitric acid (HNO_3), sulfuric acid is one of the two strong acids in acid deposition, or acid rain. The increase in acidity associated with these acids can have harmful effects on plant life, on aquatic organisms, and on human-made materials.

To try to prevent the formation of SO_2 (and subsequently H_2SO_4), one or both of two procedures are used. One procedure is to "wash" the coal first to reduce the amount of sulfur impurities. The other procedure is to install *scrubbers* in the smokestacks of power plants. The scrubbers

Acid drainage from iron mines, such as this one owned by the Rio Tinto Group, can be extremely hazardous to the environment. *(Ismael Montero Verdu/Shutterstock)*

spray the smokestack fumes with limestone ($CaCO_3$), converting SO_2 into gypsum, which can be used to manufacture wallboard, concrete, and sulfuric acid. Successful scrubbers reduce the quantity of SO_2 emissions from power plants by as much as 90 percent.

Another environmental problem associated with sulfur is called *acid mine drainage.* Groundwater in both coal and metal sulfide mines can become contaminated with sulfuric acid. This is particularly a problem in the coal mines of the Appalachian Mountains. Coal deposits often contain pyrites. Metal ores can include toxic salts of zinc, lead, arsenic, copper, and aluminum. In both cases, acid mine drainage can kill fish and other aquatic organisms, as well as corrode boats and piers. Streams sometimes turn red in color because of the iron content from the mine drainage. The problem is especially acute in the case of mines that have been abandoned, because they are not regulated in any way.

With the many compounds of sulfur found in the atmosphere, in aquatic environments, and in soils and minerals, sulfur cycles through the biosphere in much the same way that nitrogen does. However, unlike the relative abundances of nitrogen—for which the atmosphere is the major reservoir—the relative abundance of sulfur in the atmosphere is small compared with its abundance in other environments.

FROM THE ANCIENT CHINESE: GUNPOWDER AND MATCHES

Gunpowder, a remarkable substance that irreversibly changed the course of warfare, and thus world history, was invented, or perhaps simply chanced upon, in China, perhaps as far back as 2,000 years ago. According to myth or legend or oral history, a Chinese cook inadvertently mixed the correct proportions of sulfur, charcoal, and saltpeter and found it to be explosive when heated. That proportion—still in use today—is 1 part sulfur, 1.5 parts charcoal, and 7.5 parts saltpeter or potassium nitrate (KNO_3). (Sodium nitrate or "Chile saltpeter" may also be used effectively.)

As reported in a document written around 1000 C.E., the Chinese utilized the phenomenon to make fireworks and that the first firecracker was invented using bamboo stuffed with the saltpeter mixture and a wick. They are known to have tied these to arrows to be delivered to enemies in battle—a very early form of "gun." The first match was also made in China by applying a sulfur compound to a twig; the match could be lit by "touching" it to a hot coal. (Friction matches were not invented until the 1800s.)

It was not until after the Crusades were well under way in the 13th century that Europeans had their introduction to explosives. Arabian scientists had fashioned "fire arrow" projectiles to use against the infidels, and word soon was carried back to Europe, probably by Italian writer and traveler Marco Polo (ca. 1254–1354 C.E.). In fact, the

Italians were the first Europeans to find an artistic use for gunpowder; they designed beautiful fireworks for entertainment at festivals.

SULFITES AND FOOD PRESERVATION

Microorganisms such as bacteria, fungi, and molds find human foods to be favorable environments in which to thrive. As these organisms grow and multiply, their metabolic waste products contaminate food and eventually cause it to spoil. There are various ways to prevent this. One way is to freeze or refrigerate food, since microorganisms tend to be inactive at low temperatures. Another way is to dehydrate food because microorganisms cannot live in the absence of water. Still another way is to add salt to food because salt absorbs water and causes the food to become dehydrated.

There are acids that are toxic to microorganisms but harmless to humans—at least most humans. Most common are the so-called *organic acids,* which are characterized by a grouping of atoms in the form COOH. One example is benzoic acid (C_6H_5COOH), which is a naturally occurring substance and is used in several food products. Sorbic acid (C_5H_7COOH) is sometimes used, although it only inhibits the growth of molds and fungi, not bacteria. Other examples include acetic acid (CH_3COOH) and citric acid ($C_3H_5O[COOH]_3$), which are found in many natural food products—acetic acid in vinegar and citric acid in fruit.

In addition to the organic acids, there are two inorganic acids that are frequently used. Nitrous acid (HNO_2), a weak acid not to be confused with the strong acid, nitric acid (HNO_3), is formed from its salt, sodium nitrite ($NaNO_2$), and is commonly used to preserve meat products because it inhibits the growth of the microorganism responsible for *botulism.* The bisulfite ion (HSO_3^-) is a weak acid contained in the salt sodium bisulfite ($NaHSO_3$). The bisulfite ion kills bacteria in grape products and other food products, and can also be generated by exposing foods to sulfur dioxide (SO_2), in which case the bisulfite ion results from the reaction between SO_2 and H_2O. About 1 percent of the human population is allergic to ions or compounds containing sulfur, however, so foods containing bisulfite or similar forms of sulfur must be labeled accordingly.

TECHNOLOGY AND CURRENT USES

Numerous chemical manufacturing processes utilize sulfur compounds. Sulfur is used in the production of agricultural fertilizers. It is also important in the production of iron and steel, and some paint pigments contain sulfur compounds.

About 90 percent of all sulfur goes into the manufacture of sulfuric acid (H_2SO_4). Sulfuric acid is the world's most abundant industrial

A VISIT TO YELLOWSTONE

Yellowstone National Park, in the far northwest corner of Wyoming, is a hot spot for sulfur. Sitting on an ancient caldera (a type of volcanic crater) that erupts approximately every 700,000 years, the ground is slowly but continually rising at a rate of about three inches per year due to thermal expansion of the earth and rock beneath the surface. Some of the water heated underground bubbles up as hot springs and spectacular geysers that carry the characteristic sulfur dioxide aroma. Some of the water stays underground, but it is hot enough to emit steam that rises and then settles and condenses to form pretty yellow sulfur crystals around the steam vents or "fumaroles."

Researchers have recently uncovered some curious phenomena regarding sulfur-eating microbes in Yellowstone. In the Gibbon Hills and Mud Volcano regions, for example, green bacteria have been discovered that can live and propagate without air, relying only on light, hydrogen sulfide (H_2S), and CO_2 for their existence. These bacteria can give clues to the possible origin of life without an oxygen atmosphere (see chapter 5). Another experiment shows that H_2S in the park generally oxidizes to sulfur without the aid of biological systems, except at Mammoth Hot Springs, where it oxidizes biologically, with help from bacteria.

chemical, with many applications. About half of all sulfuric acid is converted into compounds used in fertilizers. A portion of the sulfur that does not go into sulfuric acid is used in the vulcanization of rubber tires, a process that strengthens tires and gives them longer lifetimes. In addition, sulfur is used in the manufacture of sulfites, used in preserving fruit and for dusting crops to control fungi. About 10 million tons (ca. 9 million tonnes) of sulfur are recovered annually from deposits in the United States. Much of it is recovered in the form of hydrogen sulfide gas (H_2S), a contaminant of natural gas, which must be removed from the natural gas. Sulfur is imported from Canada, Mexico, and

Obviously, Yellowstone is an important research area for the study of the origins of life, volcanic geology, and evolving complex systems.

Sulfur vents, such as these in Yellowstone National Park, indicate underground thermal activity. (*J. R. Douglass/Yellowstone National Park*)

Venezuela. Canada is a major exporter because it is also a major source of natural gas. Between 3 million and 5 million tons (2.7 and 4.5 million tonnes) of sulfuric acid are recycled annually from petroleum refining and chemical manufacturing processes.

Manufacturers of synthetic fibers use sulfur to *cross-link* the fibers together. Cross-linking makes the materials less flexible. In the same way, rubber manufacturers use sulfur to strengthen rubber. The addition of sulfur to rubber to make tires is called *vulcanization.* The discovery of the vulcanization process in the 1840s was a milestone in the history of invention.

Pyrite (FeS_2) is of particular interest to scientists who study the origin of life on Earth because it may provide a clue as to when oxygen became a significant portion of the atmosphere and, therefore, whether early life developed with or without it.

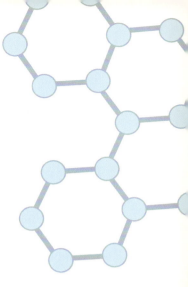

7

Selenium: Its Relevance in Health, Photocopiers, and Solar Cells

Selenium—element number 34—with a density of 4.8 g/cm$_3$ is a relatively rare element that is only the 68th most abundant element in Earth's crust. There are no selenium ores that would be profitable to mine. Selenium is usually found along with ores of sulfides, especially sulfides of copper. Canada, the United States, and Russia are the primary sources of selenium.

A general trend in the periodic table is for the sizes of atoms to increase as one descends a group of elements. The increase in size is due to each element's outermost electrons being located in progressively higher energy levels. Because atoms of nonmetals tend to be very small in size and atoms of metals tend to be relatively larger, the metallic nature of elements tends to increase toward the bottom of a column. Therefore, although sulfur and selenium form many similar

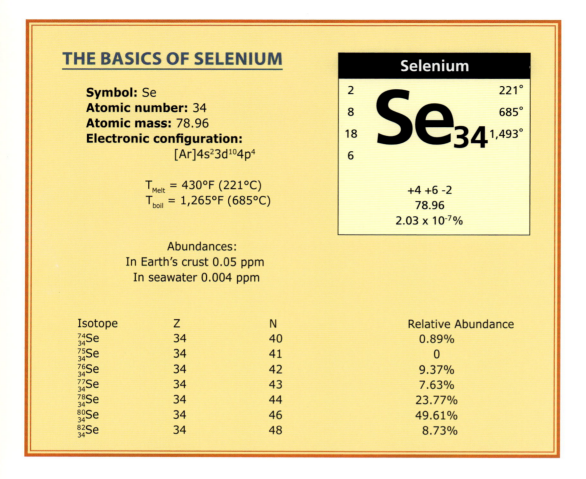

THE BASICS OF SELENIUM

Symbol: Se
Atomic number: 34
Atomic mass: 78.96
Electronic configuration:
$[Ar]4s^23d^{10}4p^4$

$T_{Melt} = 430°F (221°C)$
$T_{boil} = 1,265°F (685°C)$

Abundances:
In Earth's crust 0.05 ppm
In seawater 0.004 ppm

Selenium		
2		221°
8	**Se**	685°
18	34	1,493°
6		
	+4 +6 -2	
	78.96	
	2.03 x 10⁻⁷%	

Isotope	Z	N	Relative Abundance
$^{74}_{34}Se$	34	40	0.89%
$^{75}_{34}Se$	34	41	0
$^{76}_{34}Se$	34	42	9.37%
$^{77}_{34}Se$	34	43	7.63%
$^{78}_{34}Se$	34	44	23.77%
$^{80}_{34}Se$	34	46	49.61%
$^{82}_{34}Se$	34	48	8.73%

compounds, as the pure elements, selenium is more metallic than sulfur is. One allotrope—a gray form of selenium—is a poor conductor of electricity, but its conductor is increased by the action of light. This property makes selenium appropriate to use in devices that measure the intensity of light. A familiar example is the sensor that automatically turns house lights on at night and off in daylight.

SELENIUM IN STARS AND ON EARTH

Selenium is the first element presented in this volume that is not synthesized in stellar cores. Many elements heavier than iron slowly build up more than thousands of years in stellar atmospheres via neutron capture and electron release, with the requirement that iron 56 be available as a "seed." Because the synthesis proceeds slowly due to a low density of neutrons, it is called the "s-process." Neutrons become available from the capture of alpha-particles by ^{13}C and ^{22}Ne as follows:

$$^{13}_{6}C + {}^{4}_{2}\alpha \rightarrow {}^{16}_{8}O + {}^{1}_{0}n$$

$$^{22}_{10}Ne + {}^{4}_{2}\alpha \rightarrow {}^{25}_{12}Mg + {}^{1}_{0}n$$

Astronomers have observed mainly the Se^{3+} ion in the atmospheric spectra of old high-mass stars, called *asymptotic giant branch* (AGB) stars. These stars shed their elements into the atmosphere via thermal pulsations, which are poorly understood but are effective at providing the interstellar medium with the heavy elements now found on Earth.

In mines and on the surface of the planet, selenium is found most obviously in pyrite, where it occasionally stands in for sulfur

In mines and on the surface of the planet, selenium is found most obviously in pyrite or "fool's gold," where it occasionally stands in for sulfur, and is common in soil samples worldwide. *(Andrew Silver/USGS)*

(see chapter 6), and is common in soil samples worldwide. (Note: Poisonous amounts are rare, but this can be a problem when land is newly irrigated. The selenium formerly bound in soil dissolves and is carried downstream to wetlands, where it sits in high concentrations, endangering wildlife.) Selenium for human use is most easily available in the anthropogenic mud and flue dust created by smelting of sulfide ores; copper, silver, and lead mines provide most of the selenium needed for current technological needs.

THE DISCOVERY AND NAMING OF SELENIUM

Sulfur, selenium, and tellurium form a *triad* of elements. A triad of elements is any group of three elements that have very similar chemical and physical properties. Because their properties are so similar, these elements tend to exist together in nature. The quantities of selenium and tellurium, however, are very small compared with the quantity of sulfur found on Earth. Therefore, whereas sulfur was isolated and recognized as a pure substance in ancient times, the isolation and identification of selenium and tellurium did not occur until modern times.

One of the greatest chemists of the 19th century was the Swedish chemist Jöns Jacob Berzelius (1779–1848). In 1796, Berzelius graduated from college with a degree in medicine. However, he preferred to study chemistry, and eventually became a professor of chemistry and a member of the Swedish Academy of Sciences. Berzelius is remembered for an eclectic set of contributions to chemistry, including the following: the study of electrolytic processes; the development of our modern chemical nomenclature based on Latin names; the invention of the system for writing chemical formulas for molecules that we still use today; support for the emerging theory that matter is made of atoms; the determinations of the atomic weights of most of the elements known at that time; the discoveries of three elements—cerium, selenium, and thorium; the first theory of chemical catalysis; an explanation of isomerism; the discovery of pyruvic acid; and the coining of the word *protein*. However, Berzelius was not completely perfect; for many years he insisted that chlorine is not an element!

In the smelting of ores of metals such as copper and lead, sulfur dioxide (SO_2) is produced. Because small amounts of selenium are mixed in with the sulfur, small amounts of selenium dioxide (SeO_2)

are also produced. Since the sulfur dioxide can be collected and converted into sulfuric acid (H_2SO_4), it is easy for sulfuric acid to contain small amounts of selenium. It was from such a batch of sulfuric acid that Berzelius first isolated selenium in the year 1817. Upon reduction of the two elements to their neutral states, sulfur and selenium could be distinguished easily by their difference in color; sulfur is bright yellow and selenium is brick red in powder form. They can be separated easily by the difference in their melting points; sulfur melts at 234°F (112°C) and selenium melts at 423°F (217°C).

The name *selenium* is derived from selenium's close resemblance to tellurium, which was already known. The name *tellurium* is derived from the Latin word *tellus,* which means "earth," and the Greek word *selene* means "moon." In analogy to the proximate arrangement of Earth and its Moon, selenium and tellurium also have a very close arrangement.

THE CHEMISTRY OF SELENIUM

Like most nonmetals, selenium can exist in both positive and negative oxidation states. Selenium's compounds tend to be analogous to sulfur's compounds. Thus, compounds with similar formulas exist for both elements, as shown in the following list:

Sulfur (S) SO_2 SO_3 H_2SO_3 H_2SO_4 S_2Cl_2 SF_6 CuS

Selenium (Se) SeO_2 SeO_3 H_2SeO_3 H_2SeO_4 Se_2Cl_2 SeF_6 $CuSe$

Just as SO_2 can be a noxious component of polluted air, SeO_2 also is a very disagreeable gas; its odor has been described as that of "rotten horseradish." Selenic acid (H_2SeO_4) is very reactive; it is such a powerful oxidizing agent that it can even dissolve gold. Solid selenium is soluble in concentrated H_2SO_4. The result is a green solution of $SeSO_3$.

TOXIC BUT ESSENTIAL

Selenium is biologically important to all classes of organisms and is essential to humans; a deficiency causes serious diseases. However, an excess of selenium also causes serious diseases, including loss of hair and nails, anemia, body odor, and jaundice. Where soils are high in selenium and animals graze, the animals are in danger of accumulating sufficiently excess selenium to develop toxic reactions.

It would be unusual for humans not to obtain 70 milligrams, the minimum daily requirement, of selenium in their diet, since selenium is found in seafood, meat, and grains. Selenium has antioxidant properties as well as beneficial protein functions in the blood and certain tissues.

SELENIUM IN GLASS COLORIZING

Selenium has been found to be the most favorable element for introducing a pink color in glass products. It is a complex process, fraught with difficulties, but, when successful, delivers to the glass a distribution of spherical particles that absorb light that is not pink.

The method is based on the principle of complementary colors. All sand from which glass is made has some residual traces of iron, which impart a greenish tinge. Selenium is often used in combination with cobalt to counteract this effect to make glass clear—an imposing challenge given that the human eye can distinguish 40,000 different hues. The pervasiveness of clear glass in American society would seem to indicate that the process is simple and well understood, but that is not the case. Not only do the dosages have to be excruciatingly accurate, but also crucial are temperature, rate, and isotope control.

The dosage is difficult to control because selenium prefers to vaporize rather than mixing with the glass melt, and up to 90 percent of the selenium added can be lost. The reason is that its melting temperature of 423°F (217°C) is 31°F (17°C) higher than the temperature at which it begins to volatilize. In addition, different ionic states produce different colors; Se^{6+} and Se^{4+} are colorless, while Se^{2-} generates a reddish-brown color. Only Se^0 produces the pink needed to make clear glass. The ionic states depend critically on the oxidation state of the mixture, which is often not known. Compounds like CeO_2, AsO_2, and most sulfates oxidize selenium, which tends to bleach the pink color.

Selenium has been found to be the most favorable element for introducing a pink color in glass products. *(Laumerle | Dreamstime.com)*

Graduate theses have been written examining the chemistry, and glassmakers continue to experiment with ways to make the process more worker friendly.

TECHNOLOGY AND CURRENT USES

Selenium is biologically important to all classes of organisms and is essential to humans; a deficiency causes serious diseases. Selenium has long been lauded as an important food supplement, and livestock need a prescribed amount of selenium in grasses and weeds. An excess of selenium, however, can have dire effects, including loss of hair and

PROGRESS: REPLACED BY SILICON

Selenium has excellent light-conducting and light-converting properties. Because of its ability to convert light to electrical energy, until around 1970 it was the material of choice for rectifiers—electronic components that convert alternating current to direct current for use in a multitude of applications, including television and radio circuits. Selenium semiconductors were more efficient than the previously used copper oxide rectifiers, but could not stand the competition when cheaper silicon diodes came on the market. Now nearly all rectifiers are made with silicon components.

Still, selenium remained useful in photocopiers until around the mid-1980s, when it was recognized that even small amounts of vapor from heated copier drums impregnated with vitreous selenium had harmful health effects—mostly throat irritation. High levels of ingestion (even over a very short period) could cause vomiting and skin rashes. Organic photoconductors have been developed, which are now used almost exclusively in new photocopiers. Selenium is currently enjoying a resurgence of interest as a possibly important component in electronics for solar cells.

nails, anemia, body odor, and jaundice; hence its uses are limited in the workplace.

Photocopy machines, photocells, and light meters can use selenium as the receptor, though the vapors have caused serious problems for humans using photocopiers. Certain pigments in plastics, paints, glass, ceramics, and ink require selenium, and it is an important additive in high-temperature lubricants and steel. Selenium has been found to be the most favorable element for introducing a pink color in glass products. The coloration is a complex process, fraught with difficulties but, when successful, delivers to the glass a distribution of spherical particles that absorb all light that is not pink.

Selenium has excellent light-conducting and light-converting properties. Until the 1970s, selenium semiconductors were more efficient than the previously used copper oxide rectifiers but could not stand the competition when cheaper silicon diodes came on the market. Now nearly all rectifiers are made with silicon components. Selenium is, however, currently enjoying a resurgence of interest as a possibly important component in electronics for solar cells.

8

Conclusions and Future Directions

The periodic table of the elements is a marvelous tool, and one that scientists have only begun to investigate. The key to its utility is its organization, the patterns it weaves, and the way that it can guide the eye and the mind to understand how far science has come and where human knowledge can lead.

UNDERSTANDING PATTERNS AND PROPERTIES IN THE NONMETALS

With their understanding of the nature of the chemical bond and the quantum mechanical model of atoms and molecules, scientists now have an improved understanding of why and how metals, semimetals, and nonmetals differ from each other. The properties of nonmetals are largely determined by the number of valence electrons that nonmetallic elements have in p subshells. The patterns, or trends, in properties observed as one moves across a period of the periodic table are explained by the increasing numbers of p electrons that eventually fill a p subshell with each noble gas, beginning with neon. When p subshells fill—or when the 1s subshell fills, as in the case of helium—the isolated atoms achieve an exceptionally stable energy state, with the result that the elements known as noble gases tend not to form chemical bonds.

Along with the increase in the number of electrons comes an increase in the number of protons in the atoms' nuclei. The increase in the magnitude of positive electrical charge in the nucleus increases the electrostatic force of attraction the nucleus exerts on all of an atom's electrons, with the result that atoms tend to decrease in size as one moves across a row of the periodic table from left to right. Thus, the atoms of nonmetallic elements tend to be much smaller than the atoms of metallic elements, with helium atoms being the smallest atoms of all elements. One consequence of the smaller size is that nonmetals have little tendency to give up electrons, so that nonmetals do not conduct electricity.

The trends in properties in a family of elements observed as one descends a column of the periodic table are explained by the location of the valence electrons in successively larger energy shells, resulting in atoms that increase in size going down a column. Larger atoms can bond to a greater number of other atoms, thereby increasing the diversity of chemical compounds the elements can form. The melting and boiling points of elements also increase as the atoms become larger in size. In addition, nonmetallic elements become more metallic in nature as their atoms increase in size. In fact, the atoms of metals tend in general to be larger than the atoms of nonmetals; again, the larger sizes of metal atoms are directly responsible for the fact that metals are conductors of electricity, while nonmetals are insulators.

Carbon occupies a unique position in the periodic table. Its four valence electrons allow carbon atoms to link together in long chains, leading to the great diversity of biologically important substances, and for carbon atoms to bond to a wide variety of other elements. The relatively small size of carbon atoms makes carbon the only nonmetal in its family. Beginning with silicon and germanium, the atoms' larger sizes result in semimetallic behavior; the atoms' sizes increase sufficiently for tin and lead to display completely metallic behavior.

SPECULATIONS ON FURTHER DEVELOPMENTS

It is important for scientists to think ahead—to attempt to guess what areas are ripe for investigation and which may be bound for oblivion. If one considers recent remarkable leaps in information science, medicine, and particle physics that have materialized in the past century, it is clear that predictions are bound to be less impressive than eventualities, but there are some obvious starting points. Some of those, especially related to the nonmetals, are suggested here.

NEW PHYSICS

While it is sometimes difficult to distinguish between the sciences of physics and chemistry, there are important distinctions. The science of individual atoms and their interactions is clearly the realm of physics. The science of compound elements and their interactions is mostly designated as chemistry, though the two often meet. In the future, physicists and chemists will need to engage in extensive collaboration, which means they will need to become more familiar with what is happening in fields outside their own. Some areas, however, are firmly in the realm of physics and will require that expertise. The discipline is often difficult to quantify for practical applications, however. It can be characterized as perhaps the most basic of sciences, one that examines little-understood basic phenomena and attempts to further understanding before application.

One such phenomenon that bears further investigation is metallic hydrogen. The idea of phase transitions that do not meet the expected classifications of gas, liquid, or solid is so curious as to be considered new physics and will attract physicists of the future. An additional current hot area of research involves the newly discovered form of carbon called graphene, which exhibits high thermal conductivity and an unprecedented electron mobility. First synthesized in 2004 by researchers at the University of Manchester in England, this material—whose configuration resembles chicken wire, as would a single layer of graphite—is also extremely strong. Obvious applications in electronic circuits, which rely on electron mobility for signal speed, have made the study of graphene priority research in universities around the world.

Another study that lies in the realm of physics is the exploration of elemental nucleosynthesis in the big bang, stars, and supernovae. Astrophysics is an observational, rather than an experimental, science and is therefore relegated to a detailed long-term commitment by physics graduates.

Regarding current practical issues, physicists will be responsible for the attempt to make efficient fusion energy a reality. The process is well understood, but much more experimentation is needed for humans to realize it as a viable source of public energy.

In all areas of future scientific investigation, physicists will need to make themselves available as their expertise is needed. The collaboration between chemists and physicists may guide the future of science, industry, and the economy.

NEW CHEMISTRY

Chemistry has progressed a long way from the early years of metallurgy and alchemy. Today the chemical industry includes enterprises ranging from fossil fuels and renewable energies to biotechnology and genetic engineering, pharmaceuticals, precious and base metals, industrial gases, water purification and softening systems, agricultural fertilizers and pesticides, plastics and fibers, synthetic rubber, semiconductors, natural products derived from plants and animals, basic chemicals for industry, a wide array of household products, and radioisotopes for medical diagnosis and treatment. Add in basic and applied research activities in the government and education sectors—along with compliance with environmental regulations—and it is clear why chemistry is often called "the central science."

Modern chemical synthesis may be said to have begun with chemists' understanding of the nature of the chemical bond. The properties of matter on a molecular level—the kinds of chemical bonds in compounds, the shapes of molecules, the effects of shape on polarity and other properties—determine the properties of matter on the macroscopic, or observable, level. As chemists and other materials scientists have learned increasingly more about the properties of elements and compounds, their ability to synthesize novel and useful substances not found in nature has increased exponentially. As recently as 100 years ago, the number of known chemical compounds was perhaps a few thousand. Today, the variety of known compounds is numbered in the millions. It is a daunting effort to catalog, classify, store, and then retrieve needed information about such a huge number of substances.

Because the elements hydrogen, carbon, and oxygen are so abundant on Earth and are the building blocks of such a huge number of substances, their chemistry will always be a subject of investigation. Nitrogen and phosphorus are such important plant nutrients that agricultural scientists will continue to conduct research in finding ways to manufacture economically and in large quantities compounds that are readily accessible to plants and that promote plant growth and health.

With the ever-increasing need to develop renewable energy technologies that are economically competitive with fossil fuels—and that do not exacerbate the problems of air pollution, acid rain, or global warming—research in hydrogen-based fuel technologies will continue. For decades, materials scientists have sought catalysts that can cheaply produce hydrogen from water using only the energy in sunlight. That research will continue. Even if a large-scale transportation infrastructure based on hydrogen never proves feasible, undoubtedly there will be useful technologies that will be developed as spin-offs of the research in hydrogen.

New carbon-based compounds are produced every day in the pharmaceutical and biotechnology industries. Very few of these substances ever make it to market, but the success of products like cholesterol-lowering drugs, blood-thinners, and flu and pneumonia vaccines—just to name a few examples—demonstrates that the manufacture and production of individual drugs can be billion-dollar businesses.

Applications of fullerenes and products that employ nanotechnology are still in their infancies. There may be developments 20 or 30 years from now that no one has even thought of today. In fact, this same statement could probably be made for much of the cutting-edge research that is happening today in science and engineering. Probably no one in the 1970s would have predicted the present-day saturation of society with computers, cell phones, and a host of other electronic devices at every level of human activity. What will these devices look like in the 2030s? Certainly a great deal of miniaturization based on nanotechnology can be expected.

SI Units and Conversions

UNIT	QUANTITY	SYMBOL	CONVERSION
Base units			
meter	length	m	1 m = 3.2808 feet
kilogram	mass	kg	1 kg = 2.205 pounds
second	time	s	
ampere	electric current	A	
kelvin	thermodynamic temperature	K	1 K = 1°C = 1.8°F
candela	luminous intensity	cd	
mole	amount of substance	mol	
Supplementary Units			
radian	plane angle	rad	π / 2 rad = 90°
steradian	solid angle	sr	
Derived Units			
coulomb	quantity of electricity	C	
cubic meter	volume	m^3	1 m^3 = 1.308 $yards^3$
farad	capacitance	F	
henry	inductance	H	
hertz	frequency	Hz	
joule	energy	J	1 J = 0.2389 calories
kilogram per cubic meter	density	$kg\ m^{-3}$	1 $kg\ m^{-3}$ = 0.0624 $lb.\ ft^{-3}$
lumen	luminous flux	lm	
lux	illuminance	lx	
meter per second	speed	$m\ s^{-1}$	1 $m\ s^{-1}$ = 3.281 $ft\ s^{-1}$

UNIT	QUANTITY	SYMBOL	CONVERSION
meter per second squared	acceleration	m s^{-2}	
mole per cubic meter	concentration	mol m^{-3}	
newton	force	N	1 N = 7.218 lb. force
ohm	electric resistance	Ω	
pascal	pressure	Pa	1 Pa = 0.145 lb/in^2
radian per second	angular velocity	rad s^{-1}	
radian per second squared	angular acceleration	rad s^{-2}	
square meter	area	m^2	1 m^2 = 1.196 yards^2
tesla	magnetic flux density	T	
volt	electromotive force	V	
watt	power	W	1W = 3.412 Btu h^{-1}
weber	magnetic flux	Wb	

PREFIXES USED WITH SI UNITS		
PREFIX	SYMBOL	VALUE
atto	a	$\times 10^{-18}$
femto	f	$\times 10^{-15}$
pico	p	$\times 10^{-12}$
nano	n	$\times 10^{-9}$
micro	μ	$\times 10^{-6}$
milli	m	$\times 10^{-3}$
centi	c	$\times 10^{-2}$
deci	d	$\times 10^{-1}$
deca	da	$\times 10$
hecto	h	$\times 10^{2}$
kilo	k	$\times 10^{3}$
mega	M	$\times 10^{6}$
giga	G	$\times 10^{9}$
tera	T	$\times 10^{12}$

Prefixes attached to SI units alter their value.

List of Acronyms

ADP   adenosine diphosphate

AGB   asymptotic giant branch

ATP   adenosine triphosphate

CARB   California Air Resources Board

CNO   carbon-nitrogen-oxygen

DNA   deoxyribonucleic acid

D-T   deuterium-tritium

EPA   Environmental Protection Agency

GHG   greenhouse gas

NADPH nicotinamide adenine dinucleotide phosphate

RNA   ribonucleic acid

SUV   sport-utility vehicle

TSP   trisodium phosphate

UV   ultraviolet

VOC   volatile organic compounds

Periodic Table of the Elements

Periodic Table of the Elements

Key:

- Atomic number
- Symbol
- Atomic weight

3
Li
6.941

Legend:
- Halogens
- Metals
- Nonmetals
- Metalloids
- Unknown

1 IA																	18 VIIIA
1 **H** 1.00794	2 IIA											13 IIIA	14 IVA	15 VA	16 VIA	17 VIIA	2 **He** 4.0026
3 **Li** 6.941	4 **Be** 9.0122											5 **B** 10.81	6 **C** 12.011	7 **N** 14.0067	8 **O** 15.9994	9 **F** 18.9984	10 **Ne** 20.1798
11 **Na** 22.9898	12 **Mg** 24.3051	3 IIIB	4 IVB	5 VB	6 VIB	7 VIIB	8 VIIIB	9 VIIIB	10 VIIIB	11 IB	12 IIB	13 **Al** 26.9815	14 **Si** 28.0855	15 **P** 30.9738	16 **S** 32.067	17 **Cl** 35.4528	18 **Ar** 39.948
19 **K** 39.0938	20 **Ca** 40.078	21 **Sc** 44.9559	22 **Ti** 47.867	23 **V** 50.9415	24 **Cr** 51.9962	25 **Mn** 54.938	26 **Fe** 55.845	27 **Co** 58.9332	28 **Ni** 58.6934	29 **Cu** 63.546	30 **Zn** 65.409	31 **Ga** 69.723	32 **Ge** 72.61	33 **As** 74.9216	34 **Se** 78.96	35 **Br** 79.904	36 **Kr** 83.798
37 **Rb** 85.4678	38 **Sr** 87.62	39 **Y** 88.906	40 **Zr** 91.224	41 **Nb** 92.9064	42 **Mo** 95.94	43 **Tc** (98)	44 **Ru** 101.07	45 **Rh** 102.9055	46 **Pd** 106.42	47 **Ag** 107.8682	48 **Cd** 112.412	49 **In** 114.818	50 **Sn** 118.711	51 **Sb** 121.760	52 **Te** 127.60	53 **I** 126.9045	54 **Xe** 131.29
55 **Cs** 132.9054	56 **Ba** 137.328	57-70 ☆	72 **Hf** 178.49	73 **Ta** 180.948	74 **W** 183.84	75 **Re** 186.207	76 **Os** 190.23	77 **Ir** 192.217	78 **Pt** 195.08	79 **Au** 196.9655	80 **Hg** 200.59	81 **Tl** 204.3833	82 **Pb** 207.2	83 **Bi** 208.9804	84 **Po** (209)	85 **At** (210)	86 **Rn** (222)
87 **Fr** (223)	88 **Ra** (226)	89-102 ★	104 **Rf** (261)	105 **Db** (262)	106 **Sg** (266)	107 **Bh** (262)	108 **Hs** (263)	109 **Mt** (268)	110 **Ds** (271)	111 **Rg** (272)	112 **Cn** (277)	113 **Uut** (284)	114 **Uuq** (285)	115 **Uup** (288)	116 **Uuh** (292)		118 **Uuo** (294)

☆ Lanthanides

57 **La** 138.9055	58 **Ce** 140.115	59 **Pr** 140.908	60 **Nd** 144.24	61 **Pm** (145)	62 **Sm** 150.36	63 **Eu** 151.966	64 **Gd** 157.25	65 **Tb** 158.9253	66 **Dy** 162.500	67 **Ho** 164.9303	68 **Er** 167.26	69 **Tm** 168.9342	70 **Yb** 173.04

★ Actinides

89 **Ac** (227)	90 **Th** 232.0381	91 **Pa** 231.036	92 **U** 238.0289	93 **Np** (237)	94 **Pu** (244)	95 **Am** 243	96 **Cm** (247)	97 **Bk** (247)	98 **Cf** (251)	99 **Es** (252)	100 **Fm** (257)	101 **Md** (258)	102 **No** (259)

Numbers in parentheses are atomic mass numbers of most stable isotopes.

© Infobase Publishing

138

Table of Element Categories

Element Categories

Nonmetals
1	H	Hydrogen
6	C	Carbon
7	N	Nitrogen
8	O	Oxygen
15	P	Phosphorus
16	S	Sulfur
34	Se	Selenium

Halogens
9	F	Fluorine
17	Cl	Chlorine
35	Br	Bromine
53	I	Iodine
85	At	Astatine

Noble Gases
2	He	Helium
10	Ne	Neon
18	Ar	Argon
36	Kr	Krypton
54	Xe	Xenon
86	Ra	Radon

Metalloids
5	B	Boron
14	Si	Silicon
32	Ge	Germanium
33	As	Arsenic
51	Sb	Antimony
52	Te	Tellurium
84	Po	Polonium

Alkali Metals
3	Li	Lithium
11	Na	Sodium
19	K	Potassium
37	Rb	Rubidium
55	Cs	Cesium
87	Fr	Francium

Alkaline Earth Metals
4	Be	Beryllium
12	Mg	Magnesium
20	Ca	Calcium
38	Sr	Strontium
56	Ba	Barium
88	Ra	Radium

Post-Transition Metals
13	Al	Aluminum
31	Ga	Gallium
49	In	Indium
50	Sn	Tin
81	Tl	Thallium
82	Pb	Lead
83	Bi	Bismuth

Transactinides
104	Rf	Rutherfordium
105	Db	Dubnium
106	Sg	Seaborgium
107	Bh	Bohrium
108	Hs	Hassium
109	Mt	Meitnerium
110	Ds	Darmstadtium
111	Rg	Roentgenium
112	Cn	Copernicium
113	Uut	Ununtrium
114	Uuq	Ununquadium
115	Uup	Ununpentium
116	Uuh	Ununhexium
118	Uuo	Ununoctium

Transition Metals
21	Sc	Scandium	39	Y	Yttrium	72	Hf	Hafnium
22	Ti	Titanium	40	Zr	Zirconium	73	Ta	Tantalum
23	V	Vanadium	41	Nb	Niobium	74	W	Tungsten
24	Cr	Chromium	42	Mo	Molybdenum	75	Re	Rhenium
25	Mn	Manganese	43	Tc	Technetium	76	Os	Osmium
26	Fe	Iron	44	Ru	Ruthenium	77	Ir	Iridium
27	Co	Cobalt	45	Rh	Rhodium	78	Pt	Platinum
28	Ni	Nickel	46	Pd	Palladium	79	Au	Gold
29	Cu	Copper	47	Ag	Silver	80	Hg	Mercury
30	Zn	Zinc	48	Cd	Cadmium			

Note: The organization of periodic table of the elements, while useful to chemists and physicists, may be confusing to nonscientists in that some groupings of similar elements appear as vertical columns (halogens, for example), some as horizontal rows (lanthanides, for example), and some as a combination of both (nonmetals).

Lanthanides
57	La	Lanthanum	62	Sm	Samarium	67	Ho	Holmium
58	Ce	Cerium	63	Eu	Europium	68	Er	Erbium
59	Pr	Praseodymium	64	Gd	Gadolinium	69	Tm	Thulium
60	Nd	Neodymium	65	Tb	Terbium	70	Yb	Ytterbium
61	Pm	Promethium	66	Dy	Dysprosium	71	Lu	Lutetium

The table of element categories is intended as a quick reference sheet to easily determine which elements belong to which groups. (Element 117 does not appear in this list because it is undiscovered as of the publishing of this book.)

Actinides
89	Ac	Actinium	94	Pu	Plutonium	99	Es	Einsteinium
90	Th	Thorium	95	Am	Americium	100	Fm	Fermium
91	Pa	Protactinium	96	Cm	Curium	101	Md	Mendelevium
92	U	Uranium	97	Bk	Berkelium	102	No	Nobelium
93	Np	Neptunium	98	Cf	Californium	103	Lr	Lawrencium

Chronology

1627 Robert Boyle is born on January 25 in County Cork, Ireland.

1660 George Ernst Stahl is born on October 21 in Anspach, Bavaria.

1669 Hennig Brand of Germany discovers phosphorus.

1691 Robert Boyle dies on December 30 in London, England.

1728 Joseph Black is born on April 16 in Bordeaux, France.

1731 Henry Cavendish is born on October 10 in Nice, France.

1733 Joseph Priestley is born on March 13 in Fieldheads, Yorkshire, England.

1734 George Ernst Stahl dies on May 14 in Berlin, Germany.

1742 Carl Wilhelm Scheele is born on December 9 (or 19, exact date unknown) in Stralsund, Swedish Pomerania.

1743 Antoine-Laurent Lavoisier is born on August 26 in Paris, France.

1749 Daniel Rutherford is born on November 3 in Edinburgh, Scotland.

1756 Joseph Black prepares carbon dioxide.

1766 John Dalton is born on September 6 in Eaglesfield, Cumberland, England.

Henry Cavendish identifies hydrogen as an element.

1768 Georg Brandt dies on April 29 in Stockholm, Sweden.

1772 Daniel Rutherford discovers nitrogen.

1774 Joseph Priestley prepares oxygen.

1777 Antoine-Laurent Lavoisier explains the true nature of combustion.

1778 Joseph-Louis Gay-Lussac is born on December 6 in Saint-Leonard, France.

Sir Humphrey Davy is born on December 17 in Penzance, England.

1779 Jöns Jacob Berzelius is born on August 20 in Väversunda, Sweden.

1783 First manned balloon flights occur in France.

1786 Carl Wilhelm Scheele dies on May 21 in Köping, Västmanland, Sweden.

1794 Antoine-Laurent Lavoisier dies on the guillotine on May 8 in Paris, France.

1799 Joseph Black dies on December 6 in Edinburgh, Scotland.

1800 Friedrich Wöhler is born on July 31 in Eschersheim, Germany.

1804 Joseph Priestley dies on February 6 in Northumberland, Pennsylvania.

1808 John Dalton publishes *A New System of Chemical Philosophy*.

1810 Henry Cavendish dies on February 24 in London, England.

1817 Jöns Jacob Berzelius isolates selenium.

1827 Friedrich Wöhler synthesizes urea.

1834 Dmitri Mendeleev is born on February 8 in Tobolsk, Siberia.

1844 John Dalton dies on July 27 in Manchester, England.

1848 Jöns Jacob Berzelius dies on August 7 in Stockholm, Sweden.

1856 Joseph John ("J. J.") Thomson is born on December 18 in Manchester, England.

1858 Max Planck is born on April 23 in Kiel, Germany.

1868 Fritz Haber is born on December 9 in Breslau, Silesia (now part of the Czech Republic).

1869 Dmitri Mendeleev presents his proposal for the periodic table of the elements, March 6.

1871 Ernest Rutherford is born on August 30 near Nelson, New Zealand.

1874 Carl Bosch is born on August 27 in Cologne, Germany.

1882 Friedrich Wöhler dies on September 23 in Göttingen, Germany.

1885 Niels Bohr is born on October 7 in Copenhagen, Denmark.

1887 Erwin Schrödinger is born on August 12 in Vienna, Austria.

1892 Louis de Broglie is born August 15 in Dieppe, Seine-Maritime, France.

1893 Harold Urey is born on April 29 in Walkerton, Indiana.

1896 Henri Becquerel discovers radioactivity.

1897 J. J. Thomson discovers the electron.

1900 Max Planck proposes dual nature of light.

1907 Dmitri Mendeleev dies on February 2 in St. Petersburg, Russia.

1911 Ernest Rutherford's research leads to the nuclear model of the atom.

1913 Bohr proposes the planetary model of the atom.

Fritz Haber and Carl Bosch independently develop the Haber-Bosch process for manufacturing ammonia.

1923 Louis de Broglie proposes wave nature of matter.

1926 Erwin Schrödinger proposes wave mechanical model of the atom.

1927 Clinton Davisson and Lester Germer demonstrate electron diffraction at the Bell Laboratories in New Jersey.

1931 Harold Urey discovers heavy water.

1934 Fritz Haber dies on January 29 in Basel, Switzerland.

1937 The *Hindenburg* explodes on May 6 in New Jersey.

1940 Carl Bosch dies on April 26 in Heidelberg, Germany.

J. J. Thomson dies on August 30 in Cambridge, England.

1947 Max Planck dies on October 4 in Göttingen, Germany.

1948 A killer smog episode occurs on October 26–31 in Donora, Pennsylvania.

1952 A killer smog episode occurs on December 5–9 in London, England.

1961 Erwin Schrödinger dies on January 4 in Vienna, Austria.

1962 Niels Bohr dies on November 18 in Copenhagen, Denmark.

1970 The United States Environmental Protection Agency is created.

The United States Congress passes the Clean Air Act.

1981 Harold Urey dies on January 5 in La Jolla, California.

1985 Harold Kroto, Robert Curl, and Richard Smalley discover fullerenes.

1987 Louis de Broglie dies on March 19 in Louveciennes, near Paris, France.

1997 Representatives of nations around the world sign the Kyoto Protocol to reduce greenhouse gas emissions.

2005 Richard Smalley dies on October 28.

2007 Former U.S. vice president Al Gore shares the Nobel Peace Prize with the Intergovernmental Panel on Climate Change for their efforts to warn of the effects of humanmade climate change.

2009 U.S. president Barack Obama selects Nobel Prize–winning physicist Steven Chu as energy secretary.

Glossary

absorption spectrum   a spectrum in which some frequencies have been absorbed; in the visible region, absorption produces a pattern of dark lines against a background of colors.

acid   a type of compound that contains hydrogen and dissociates in water to produce hydrogen ions.

acid-base conjugate pair   two chemical entities that differ only in one hydrogen ion. Removing a hydrogen ion from an acid converts it into the conjugate base; adding a hydrogen ion to a base converts it into its conjugate acid.

acid deposition   any acid precipitated out of the atmosphere in the form of rain, snow, sleet, hail, aerosols, or dust.

acidic   referring to a solution with a pH less than 7; a solution with a relatively high concentration of hydrogen ions.

acid mine drainage   a kind of pollution in which bacteria convert sulfur impurities in coal (in mines) into sulfuric acid.

acid rain   any acid—although usually H_2SO_4 or HNO_3—precipitated out of the atmosphere in the form of rain.

actinide   the elements ranging from thorium (atomic number 90) to lawrencium (number 103); they all have two outer electrons in the "7s" subshell plus increasingly more electrons in the "5f" subshell.

adenosine triphosphate   an organic compound that is important as a transporter of chemical energy in living organisms, as, for example, in muscle contraction.

alchemy   the precursor of modern chemistry; practitioners investigated the properties of chemical substances and the transformations of substances into other substances.

alcohol   one of a group of organic compounds characterized by a –C–OH group.

aldehyde   one of a group of organic compounds characterized by a –C=O group; at least one hydrogen atom must be bonded to the carbon.

algal bloom   a proliferation of algae in natural aquatic ecosystems.

alkali metal   the elements in the first column of the periodic table (exclusive of hydrogen); they all are characterized by a single valence electron in an "s" subshell.

alkaline   *See* **basic**.

alkaline earth   the elements in the second column of the periodic table; they all are characterized by two valence electrons that fill an "s" subshell.

alkane   a hydrocarbon of the general formula C_nH_{2n+2}; the carbon atoms are all connected to other carbon atoms by single bonds.

alkene   a hydrocarbon of the general formula C_nH_{2n}; at least two carbon atoms are connected together with a double bond.

alkyne   a hydrocarbon of the general formula C_nH_{2n-2}; at least two carbon atoms are connected together with a triple bond.

allotropic forms (or allotropes)   two or more different forms of an element; the difference could be due to different numbers or arrangements of the atoms in molecules or due to different crystal structures.

alpha decay   a mode of radioactive decay in which an alpha particle—a nucleus of helium 4—is emitted; the daughter isotope has an atomic number two units less than the atomic number of the parent isotope, and a mass number that is four units less.

amine   one of a group of organic compounds derived from ammonia (NH_3) by replacing one or more hydrogen atoms with a hydrocarbon group.

amphiprotic   describing a substance that can act as both an acid and a base.

anesthetic   a substance that reduces sensitivity to pain.

angular momentum   the product of the mass, the velocity, and the radius of curvature of an object that orbits some central object.

angular velocity   the product of the velocity and the radius of curvature of an object that orbits some central object.

anion   an atom with one or more extra electrons, giving it a net negative charge.

annihilation   the process by which a particle and its antiparticle come into contact with each other and destroy each other; the conversion of the mass and kinetic energies of the colliding particles into pure energy, usually in the form of gamma rays.

anode   the electrode in an electrochemical cell at which oxidation occurs.

antimatter   the opposite of matter, made up of antiparticles.

antioxidant   any substance that reduces damage to living tissues due to oxygen such as that caused by free radicals; free radicals are highly reactive chemicals that attack molecules by capturing electrons and thus modifying chemical structures.

antiparticle   a subatomic particle that has the same mass as another particle and equal but opposite values of some other property or property, e.g., its electrical charge.

aqueous   describing a solution in water.

asymptotic giant branch (AGB)   an area of the Hertzsprung-Russell diagram above the main sequence line, where some high mass stars (AGB stars) are mapped for luminosity and temperature.

atmosphere   the layers of gases that surround Earth and are attracted to Earth by the force of gravity; a unit of pressure (abbreviated "atm": 1 atm = 760 mmHg).

atom   the smallest part of an element that retains the element's chemical properties; atoms consist of protons, neutrons, and electrons.

atomic mass   the mass of a given isotope of an element—the combined masses of all its protons, neutrons, and electrons.

atomic number   the number of protons in an atom of an element.

atomic weight   the mean weight of the atomic masses of all the atoms of an element found in a given sample, weighted by isotopic abundance.

ATP   *See* **adenosine triphosphate**.

azotobacter   *See* **nitrification**.

base   a substance that reacts with an acid to give water and a salt; a substance that, when dissolved in water, produces hydroxide ions.

basic   referring to a solution with a pH greater than 7; a solution with a relatively low concentration of hydrogen ions.

beta decay   a mode of radioactive decay in which a beta particle—an ordinary electron—is emitted; the daughter isotope has an atomic number one unit greater than the atomic number of the parent isotope, but the same mass number.

big bang theory   a theory of cosmology in which the expansion of the universe began with a primeval explosion.

biogeochemical cycle   the processes by which atoms are cycled in the biosphere.

biomass   any accumulation of organic material produced by living organisms.

biosphere   the whole of Earth's surface, the sea, and the atmosphere that is inhabited by living organisms.

botulism   a serious illness caused by a nerve toxin that is produced by the bacterium *Clostridium botulinum*; causes paralysis. Most commonly caused by eating foods that contain the botulism toxin.

branched chain   a grouping of atoms in which side chains are connected to a main chain.

brimstone   an archaic reference to sulfur.

buffer   a solution that resists changes in pH when an acid or base is added or when the solution is diluted.

calcination   the formation of calcium carbonate from an aqueous solution containing calcium and carbonate ions.

calx   the oxide of a metal formed by heating the metal in air.

carbohydrate   literally, "carbon plus water"; a bio-organic compound that contains the elements carbon, hydrogen, and oxygen; the main classes of carbohydrates are sugars, starches, and cellulose.

carboxylic acid   one of a group of organic compounds characterized by a –COOH group;

carnivore   an animal that eats other animals.

catalyst   a chemical substance that speeds up a chemical reaction without itself being consumed by the reaction.

cathode   the electrode in an electrochemical cell at which reduction takes place.

cation   an atom that has lost one or more electrons to acquire a net positive charge.

Celsius (°C)   the temperature scale on which the freezing point of water at 1 atm pressure is defined as 0°C and the boiling point of water at 1 atm pressure as 100°C.

chemical bond   a strong electrostatic attraction between atoms within a molecule or in a crystalline solid.

chemical change   a change in which one or more chemical elements or compounds form new compounds; in a chemical change, the names of the compounds change.

chloroplast   the part of a plant cell that contains chlorophyll; the part of a plant cell where photosynthesis occurs.

chromosome   the thin structures in plant and animal cells that carry the genes.

CNO cycle   CNO = carbon, nitrogen, and oxygen; refers to the stellar nuclear fusion sequence in which nitrogen and oxygen nuclei are created from carbon nuclei.

complex (ion)   any ion that contains more than one atom. *See* **poly-atomic**.

complex carbohydrate   one of a group of large organic compounds that consist of many simple sugars that have been connected together; referring usually to starches and cellulose.

compound   a pure chemical substance consisting of two or more elements in fixed, or definite, proportions.

conductor   *See* **electrical conductor**.

conjugate pair   *See* **acid-base conjugate pair**.

cosmic ray   particles that enter Earth's atmosphere from outer space.

covalent bond   a chemical bond formed by sharing valence electrons between two atoms (in contrast to an ionic bond in which one or more valence electrons are transferred from one atom to another atom).

critical point temperature   the temperature of a pure substance at which the liquid and gaseous phases become indistinguishable.

cryogenics   the production of very low temperatures and the study of the properties of materials at low temperatures.

cyclic compound   a compound in which at least one group of atoms is connected together in a closed ring.

daughter isotope   the product of a radioactive decay event; what is formed when a parent isotope decays.

decomposer   an organism that obtains nutrients by feeding on dead organisms or plant or animal wastes.

dephlogisticated air   air that does not contain phlogiston.

deuterium   the isotope of hydrogen that has a mass number of 2 (because it contains both a proton and a neutron in its nucleus); also called "heavy hydrogen."

deuteron   the nucleus of a deuterium atom.

dissociation   the event of a molecule breaking apart into ions.

distillation   the process by which a liquid is boiled and the vapor then collected and allowed to condense back into the liquid again; often used as a means of purifying a liquid by leaving behind nonvolatile impurities.

doubly excited state   an excited state of an atom or ion in which two electrons are in close enough proximity that their probability distributions, or wave functions, overlap.

ductility   the ability of certain metals to be able to be drawn into thin wires without breaking.

elastomer   a natural or synthetic rubberlike substance; a substance that can re-form its original shape after having been deformed.

electrical conductivity   the ability of a metal or other substance to conduct an electrical current.

electrolysis   the process that causes a chemical reaction to take place because of the energy supplied by an electrical current.

electromagnetic radiation   the stream of photons associated with electric and magnetic fields; visible electromagnetic radiation is usually called "light."

electron   a negatively charged subatomic particle—a component in all atoms.

electron affinity   the energy released when a neutral atom gains an extra electron to form a negative ion.

electron correlation   the interaction between electrons in an atom or molecule due to the electrical repulsion between the electrons.

electronegativity   the relative tendency of the atoms of an element to attract the electrons shared between two atoms in a chemical bond.

electronic configuration   a description of the arrangement of the electrons in an atom or ion showing the numbers of electrons occupying each subshell.

electrostatic   the type of interaction that exists between electrically charged particles; electrostatic forces attract particles together if the particles have charges of opposite sign, while the forces cause particles that have charges of like sign to repel each other.

element   a pure chemical substance that contains only one kind of atom.

elixir   the "philosopher's stone"; a hypothetical substance capable of transforming metals of lesser value into precious metals.

emission spectrum   the pattern of light (or other regions of the electromagnetic spectrum) produced when electrons in atoms make a transition from a subshell of higher energy state to a subshell of lower energy state.

empirical   a result that is obtained from experiment or observation instead of being deduced from theory.

endothermic   a chemical reaction in which heat is absorbed from the surroundings.

energy   the measure of an object or system's ability to do work on other objects or systems in its surroundings.

enzyme   a biological catalyst; one of many kinds of protein molecules present in living organisms that speed up the biochemical reactions taking place in the cells and tissues of those organisms.

ester   one of a group of organic compounds that is derived from the reaction between a carboxylic acid and an alcohol.

ether   one of a group of organic compounds that contains two carbon atoms linked together by an oxygen atom.

eutrophication   the process by which a body of water—usually a lake—builds up an excess amount of organic matter through the process of photosynthesis.

excited state   a state of an elementary particle such as an electron in which it has more energy than it would in its lowest energy ("ground") state.

exothermic   a chemical reaction that liberates heat to the surroundings.

extranuclear   located outside the nucleus of an atom; electrons are extranuclear subatomic particles.

fallout   radioactive particles deposited from the atmosphere from either a nuclear explosion or a nuclear accident.

family   *See* **group**.

femto   the metric prefix that specifies 10^{-15} of a unit; symbol = f.

fermentation   the series of biochemical reactions in which carbohydrates are converted into ethanol and carbon dioxide.

fission   *See* **nuclear fission**.

fix   *See* **nitrification**.

flash point   the temperature at which the vapor above a volatile liquid is capable of burning with a brief flash when exposed to a flame.

fluorescence   the spontaneous emission of light from atoms or molecules when electrons make transitions from states of higher energy to states of lower energy.

fossil fuel   the remains of decomposed organic matter that have been converted into coal, petroleum, and natural gas.

free radical   an atom or group of atoms that contain one or more unpaired electrons.

frequency   the rate at which an event repeats itself at regular intervals of time; in wave motion, frequency is the number of waves, or cycles, per second.

fuel cell   an electrochemical cell in which chemical energy is converted directly into electrical energy.

fullerene   an allotrope of carbon made by joining together five- and six-membered rings of carbon atoms.

fumarole   a hole in the vicinity of a volcano from which smoke and gases issue.

functional group   a group of atoms within a molecule that determines the characteristic chemical and physical properties of that substance.

fused rings   two rings of molecules that share a common side rather than being linked together between two atoms; several rings could be linked together in this fashion.

fusion   *See* **nuclear fusion**.

gamma decay   a mode of radioactive decay in which a very high energy photon of electromagnetic radiation—a gamma ray—is emitted; the daughter isotope has the same atomic number and mass number as the parent isotope, but lower energy.

gene   the basic unit of heredity; the fundamental part of a chromosome.

geodesic   the shortest distance between two points on a curved surface; for example, the shortest distance between two points on Earth's surface.

giga   metric prefix meaning 10^9, or one billion; symbol = G.

GPa   abbreviation for "gigapascal," or one billion pascals, a unit of pressure.

greenhouse effect   the warming of Earth's atmosphere by the absorption of outgoing infrared radiation.

greenhouse gas   an atmospheric gas that absorbs infrared radiation emitted by Earth and that contributes to warming of the atmosphere.

group   the elements that are located in the same column of the periodic table; also called a family, elements in the same column have similar chemical and physical properties.

guano   deposits of the excrement of organisms such as sea birds and bats; used as fertilizer because of its high nutrient content.

Haber-Bosch process   the large-scale industrial process in which ammonia is synthesized from hydrogen and nitrogen gas in the presence of a catalyst.

half-life   the time required for half of the original nuclei in a sample to decay; during each half-life, half of the nuclei that were present at the beginning of that period will decay.

halogen   the elements in column VIIB of the periodic table; all of them share a common set of seven valence electrons in an nth energy level such that their outermost electronic configuration is "ns^2np^5."

herbivore   an animal that eats plants only.

Hertzsprung-Russell (HR) diagram   used in astrophysics, a graph that plots luminosity versus surface temperature of a star.

hydride   a compound containing hydrogen; the negative ion of hydrogen, symbolized as H^-.

hydrocarbon   a chemical compound that contains only hydrogen and carbon atoms.

hydrocracking   the industrial process of breaking larger hydrocarbons into smaller compounds in the presence of hydrogen gas and a catalyst.

hydrodealkylation   the process of adding hydrogen atoms to a more complex hydrocarbon to remove a functional group, thereby converting it into a simpler hydrocarbon.

hydrodesulfurization   in the petroleum industry, the process of removing sulfur from gasoline by the addition of hydrogen gas to form hydrogen sulfide, H_2S.

hydrogenation   the addition of hydrogen atoms to organic compounds to convert multiple bonds into single bonds; most commonly the conversion of oils into fats.

hydrogen bond   a "bridge," or strong electrostatic attraction between two molecules in which a hydrogen atom is shared by the two molecules; the hydrogen atom must be bonded to an atom of nitrogen, oxygen, or fluorine.

hydrosphere   the watery part of Earth's environment; includes the oceans and all freshwater bodies.

hyperbaric chamber   a pressurized room or vessel in which a person can breathe a higher-than-normal concentration of oxygen gas.

hypoxemia   a medical condition in which there is a deficiency of oxygen in the arteries.

hypoxia   a medical condition in which there is a shortage of oxygen in the blood.

incandescence   the emission of light by a substance that results from the substance being heated to a high temperature.

icosahedron   in geometry, a regular convex polygon with 20 faces.

immiscible   a description of two liquids that do not mix together, usually because of a noticeable difference in the polarities of the two liquids; the opposite of miscible.

index of refraction   a measure of the degree light will bend upon entering a material

inert   an element that has little or no tendency to form chemical bonds; the inert gases are also called noble gases.

infrared   the region of the electromagnetic spectrum just lower in energy than the visible region; infrared radiation has relatively low frequencies and long wavelengths, and is invisible to humans.

inorganic compound   any pure substance that does not contain at least the elements carbon and hydrogen. Carbon dioxide (CO_2) and carbon monoxide (CO) do not satisfy the definition of an organic compound because they do not contain hydrogen; otherwise, an inorganic compound is any compound that does not contain carbon.

insulator   a material that is a poor conductor of electricity.

intermolecular force   an electrostatic force that attracts two molecules together; the molecules can be the same compound or molecules of different compounds; for example, attractive forces that hold molecules together in a liquid as opposed to the molecules escaping into the vapor phase.

intramolecular force   an electrostatic force that attracts two parts of a molecule together; for example, the hydrogen bonding that holds together the two strands of a DNA molecule.

ion   an atom or group of atoms that have a net electrical charge.

ionic bond   a strong electrostatic attraction between a positive ion and a negative ion that holds the two ions together.

isomers   two or more chemical compounds that have the same elements and the same number of each kind of atom but in which the atoms are arranged in structures that differ in the manner in which the atoms are connected together.

isotope   a form of an element characterized by a specific mass number; the different isotopes of an element have the same number of protons but different numbers of neutrons, hence different mass numbers.

joule   a unit of energy named in honor of the physicist James Joule; a joule is the same as the combination of units $kg \cdot m^2/s^2$; symbol = J.

Kelvin (K)   the unit of temperature on the absolute temperature scale based on the Celsius temperature scale; 0 kelvin = –273.16°C.

ketone   one of a group of organic compounds that contains a –C=O group; the carbon atom is also bonded to two hydrocarbon groups.

kinetic energy   the energy an object has due to its motion; kinetic energy = $(1/2)mv^2$, where m = the object's mass and v = the object's speed.

lanthanide   the elements ranging from cerium (atomic number 58) to lutetium (number 71); they all have two outer electrons in the "6s" subshell plus increasingly more electrons in the "4f" subshell.

law of buoyancy   the law of physics that states that the buoyant force on an object that is submerged, or partially submerged, in a fluid is equal to the weight of the fluid the object displaces.

limiting factor   in an ecosystem, a necessary resource that is primarily responsible for limiting the population size of a species.

lipid   one of a group of organic compounds consisting mostly of carbon and hydrogen; lipids have relatively high molecular weights and are insoluble in water.

lithosphere   the solid part of Earth: the crust, mantle, and core.

litmus   a natural dye that changes color depending upon the acidity of a solution; litmus is blue in acid solution and red in basic solution.

luster   the shininess associated with the surfaces of most metals.

macromolecule   a large molecule, usually containing hundreds of atoms.

main group element   an element in one of the first two columns or one of the right-hand six columns of the periodic table; distinguished from transition metals, which are located in the middle of the table, and rare earths, which are located in the lower two rows shown apart from the rest of the table.

malleability   the ability of a substance such as a metal to change shape without breaking; metals that are malleable can be hammered into thin sheets.

mass   a measure of an object's resistance to acceleration; determined by the sum of the elementary particles composing the object.

mass number   the sum of the number of protons and neutrons in the nucleus of an atom. *See* **isotope**.

matter   an object that possesses mass both at rest and in motion.

melanoma   a serious form of skin cancer resulting from too much exposure to ultraviolet radiation.

mercaptan   one of a group of organic compounds that contain an –SH group; compounds in this group are known for their strong, disagreeable odor.

metabolic rate   in humans, the amount of energy expended either during rest or during some activity.

metal   any of the elements characterized by being good conductors of electricity and heat in the solid state; approximately 75 percent of the elements are metals.

metalloid (also called semimetal)   any of the elements intermediate in properties between the metals and nonmetals; the elements in the periodic table located between the metals and nonmetals.

metallurgy   the technology of producing metals from their ores, purifying the metals, and fashioning alloys from them.

metamorphic the result of change; metamorphic rocks were originally igneous or sedimentary rocks that changed into other rocks
usually because of high pressures or temperatures.

metastable an unstable energy state that is able to persist for some
period of time; eventually an atom or molecule in a metastable state
will lose enough energy to make the transition to a stable state of
lower energy.

miscible a description of two liquids that can mix together in all proportions, usually due to similar polarities; the opposite of immiscible.

moderator in a nuclear reactor, a medium that slows the speed of initially fast-moving neutrons.

monoclinic a form of crystal structure derived from a simple cube;
the three sides are changed to being unequal in length, and one
angle between sides is changed to a value different from 90°.

nano the metric prefix that specifies one billionth of a unit; symbol
= n.

nanofiber a fiber in which the cross-section has a diameter of less
than 100 nanometers.

nanofoam an allotrope of carbon in which carbon tendrils exist in a
low-density configuration.

nanotube a structure of carbon atoms in which the atoms are rolled
into tubelike structures that are only one atom thick.

narcosis the condition of an organism characterized by a chemical-
induced unconsciousness or immobility.

natural gas a mixture of gases associated with petroleum deposits;
mostly consisting of methane gas, with smaller quantities of ethane,
propane, and butane.

neutralization a chemical reaction in which an acid and a base combine to form a solution with a neutral pH. *See* **pH.**

neutron the electrically neutral particle found in the nuclei of atoms.

nitrification the conversion of ammonia or ammonium ions into
nitrite ions and then nitrate ions to make the nitrogen accessible
for assimilation by plants; this process is made possible through the

activity of nitrogen-fixing bacteria found in the root nodules of certain classes of plants.

nitrogen-fixing bacteria   *See* **nitrification**.

noble gas   any of the elements located in the last column of the periodic table—usually labeled column VIII or possibly column 0.

nonmetal   the elements on the far right-hand side of the periodic table that are characterized by little or no electrical or thermal conductivity, a dull appearance, and brittleness.

nonpolar   referring to a molecule in which the distribution of electrical charges is completely symmetrical; the opposite of **polar**.

nuclear fission   the process in which certain isotopes of relatively heavy atoms such as uranium or plutonium break apart into fragments of comparable size; accompanied by the release of large amounts of energy.

nuclear fusion   the process in which certain isotopes of relatively light atoms such as hydrogen or helium can combine to form heavier isotopes; accompanied by the release of large amounts of energy.

nucleic acid   an acidic group of atoms found in the nuclei of cells, specifically in molecules of DNA or RNA.

nucleon   the particles found in the nuclei of atoms. *See* **neutron** and **proton**.

nucleus   the small, central core of an atom.

omega (Ω)   the letter of the Greek alphabet used as the symbol for the unit of electrical resistance, the ohm.

organic acid   organic compounds that contain the functional group –COOH.

organic compound   a pure chemical substance that contains carbon and hydrogen and possibly other elements.

organophosphate   an organic compound that contains a phosphate group; usually refers to a class of chemical insecticides.

orthorhombic   (sometimes **rhombic**) a form of crystal structure derived from a simple cube; the three sides are changed to being unequal in length while the angles between sides all remain at 90°.

oxidation   an increase in an atom's oxidation state; accomplished by a loss of electrons or an increase in the number of chemical bonds to atoms of other elements. *See* **oxidation state**.

oxidation state   a description of the number of atoms of other elements to which an atom is bonded. A neutral atom or neutral group of atoms of a single element is defined to be in the zero oxidation state. Otherwise, in compounds, an atom is defined as being in a positive or negative oxidation state, depending upon whether the atom is bonded to elements that, respectively, are more or less electronegative than that atom is.

parent isotope   an atom that undergoes radioactive decay; the new atom that is formed is called the daughter isotope.

period   any of the rows of the periodic table; rows are referred to as periods because of the periodic, or repetitive, trends in properties of the elements.

periodic table   an arrangement of the chemical elements into rows and columns such that the elements are in order of increasing atomic number, and elements located in the same column have similar chemical and physical properties.

persistent   refers to toxic or harmful chemicals unnatural to the environment that tend to stay in the environment for long periods of time without breaking down into harmless substances.

pH   a measure of the degree of acidity of a solution; acid solutions have low pH's (less than 7), basic (or alkaline) solutions have high pH's (greater than 7), and neutral solutions have pH's of 7.

philosopher's stone   a mysterious unknown substance sought by medieval alchemists that was believed would be capable of transmuting common metals into gold.

phlogiston   a colorless, odorless, and weightless substance that was believed during the 17th and 18th centuries to be present in a flammable substance and that was released to the atmosphere during combustion; in addition, phlogiston was believed to be present in metals and was released when metals rusted.

phosphorescence   chemical luminescence, brought about by the reaction of phosphorus with oxygen.

photochemical   *See* **photochemistry**.

photochemical smog   more properly called "photochemical haze," a combination of air pollutions consisting mostly of oxides of nitrogen, ozone, carbon monoxide, and hydrocarbons.

photochemistry   the sub-discipline of chemistry concerned with reactions initiated by the absorption of electromagnetic radiation.

photon   the particle of light or quantum of energy of electromagnetic radiation.

photosynthesis   the chemical process by which a plant converts carbon dioxide and water into simple organic compounds (e.g., sugars) and oxygen gas.

physical change   any transformation that results in a substance's physical state, color, temperature, dimensions, or other physical properties; the chemical identity of the substance remains unchanged in the process.

physical state   the condition of a chemical substance being either a solid, liquid, or gas.

phytochemical   a chemical compound beneficial to human health that is derived from plants.

Planck's constant   in quantum theory, the proportionality constant that relates the energy of photons to their frequency.

polar   referring to a molecule in which the distribution of electrical charges is unsymmetrical so that one side of the molecule carries a slightly positive charge and the other side of the molecule carries a slightly negative charge.

polarity   a condition of a molecule in which there is an uneven distribution of electrical charge among the atoms; one end of the molecule has a net positive charge, while the other end has a net negative charge.

polyatomic   a molecule or ion that contains two or more atoms.

polymer   a large molecule consisting of hundreds or thousands of smaller units linked together in a chain.

polyphosphate   a polymer consisting of two or more phosphate groups linked together in a chain.

post-transition metal   the metallic elements found in the periodic table to the right of the transition metals, including aluminum; the metals located in the *p*-block of the table.

potential energy   the energy an object possesses by virtue of its position in a force field; common examples are gravitational and electrical potential energy.

ppm   abbreviation for "parts per million"; used to express concentrations of 1 gram per million grams of a sample or 1 particle per million particles.

product   the compounds that are formed as the result of a chemical reaction.

productivity   a measure of the rate of plant growth in aquatic ecosystems.

protein   bio-organic compounds formed by linking together amino acids in a chain.

proton   the positively charged subatomic particle found in the nuclei of atoms.

quantum   a unit of discrete energy on the scale of single atoms, molecules, or photons of light (plural = **quanta**).

quantum theory   the branch of physics incorporating the dual nature of matter and light that describes matter and energy on a submicroscopic level.

radiation   energy due to waves or moving subatomic particles.

radioactive   the description of a subatomic particle that is capable of undergoing radioactive decay.

radioactive decay   the disintegration of a nucleus accompanied by the emission of a subatomic particle or gamma ray.

radioactivity   *See* **radioactive**.

rare earth element   the metallic elements found in the two bottom rows of the periodic table; the chemistry of their ions is determined by electronic configurations with partially filled *f* subshells. *See* **actinide** and **lanthanide**.

reactant   the chemical species present at the beginning of a chemical reaction that arrange atoms to form new species.

rectifier   a component of an electrical circuit that converts alternating current into direct current.

redshift   an increase in wavelength of electromagnetic radiation; the color of light emitted by a source that is moving away from an observer is shifted toward the red end of the electromagnetic spectrum.

redshift factor   the factor by which the universe has expanded since the light we observe from a distant galaxy left that galaxy.

reduction   a decrease in an atom's oxidation state; accomplished by a gain of electrons or a decrease in the number of chemical bonds to atoms of other elements. *See* **oxidation state**.

residence time   the average amount of time that a molecule of a particular substance spends in the atmosphere.

residue   the molecular fragment of an amino acid that determines the identity of that amino acid.

resistivity   the intrinsic resistance a material has to the flow of electrical current.

resonant   condition of a natural frequency of vibration characterized by a comparatively large amplitude.

rhombohedral   a form of crystal structure derived from a simple cube; the three sides all have equal length but the angles between sides have values different from $90°$.

saturated   referring to an organic compound in which all of the carbon-carbon bonds are single bonds.

scrubber   an industrial device that removes air pollutants from chimneys or smokestacks.

sedimentary   rocks that result from the accumulation and compaction of sediments; mostly formed by the breaking down of larger rocks into small particles that form a deposit.

sequestration   the long-term storage of a chemical substance; usually refers to storing the carbon dioxide that results from the combustion of fossil fuels rather than allowing it to accumulate in the atmosphere.

shell   all of the orbitals that have the same value of the principal energy level, n.

smog   literally, "smoke + fog"; air that is polluted due to the presence of combustion products such as particulate matter, carbon monoxide, and sulfur dioxide.

solution   A homogeneous mixture of two or more substances.

solvent   a medium in which other substances can be dissolved to make a solution; solvents are usually, but not necessarily, liquids.

spectroscope   an instrument that produces a spectrum for visual analysis.

spectroscopy   the analysis of spectra for chemical analysis, used for finding the elemental composition of celestial objects and for determining atomic and molecular energy levels.

spectrum   the range of frequencies of electromagnetic radiation obtained when the radiation is dispersed.

spin   the intrinsic angular momentum of a particle, atom, or nucleus.

spontaneous fission   the fission of a nucleus without the event being initiated by the absorption of a neutron.

standing wave   a wave that is reflected back on itself to match peaks and valleys because the length of the medium perfectly corresponds to an integer number of wavelengths.

steroid   one of a group of organic compounds that contains four saturated fused rings; examples include sex hormones, bile acids, and vitamin D.

straight chain   in organic compounds, the arrangement of carbon atoms into a linear chain as opposed to a ring or a chain that has side branches.

stratosphere   the layer of the atmosphere lying above the troposphere; at the equator, it extends to an altitude of about 80 miles (128.8 km).

strong acid   an acid that dissociates completely into ions when added to water.

strong force   the force that binds together the particles inside an atom's nucleus.

subatomic   the particles that make up an atom.

sublimation   the change of physical state in which a substance goes directly from the solid to the gas without passing through a liquid state.

sublimation point temperature   the temperature of a pure substance at which the solid and gaseous phases become indistinguishable. Only carbon and arsenic are capable of sublimation.

subshell   all of the orbitals of a principal shell that lie at the same energy level.

sugar   one of a group of water-soluble carbohydrates that have low molecular weight and that usually have a sweet taste.

supernova   a stellar explosion in which a star suddenly increases its luminosity by an order of magnitude in the thousands or millions, and is completely blown apart.

surface tension   the force due to intermolecular attractions that tends to attract molecules at the surface of a liquid back toward the interior.

symbiotic relationship   the living together or in close proximity to each other of two or more organisms of different species.

synchrotron radiation   radiation emitted when charged particles travel in a curved path.

thermal conductivity   a measure of the ability of a substance to conduct heat.

thermal neutron   a neutron that is moving at a relatively slow speed.

tonne   a metric ton, equivalent to 1,000 kilograms.

transactinide   Any element with atomic number 104 or higher.

transition metal   the metallic elements found in the 10 middle columns of the periodic table to the right of the alkaline earth metals; the chemistry of their ions is determined by electronic configurations with partially filled *d* subshells.

transition rate   the rate at which an electron in an atom jumps from one energy level to another energy level.

transmutation   conversion by way of a nuclear reaction of one element into another element; in transmutation, the atomic number of the element must change.

transuranium element   any element in the periodic table with an atomic number greater than 92.

triad   any group of three elements that exhibit very similar chemical and physical properties.

triple alpha process   a series of two nuclear reactions in which three alpha particles are fused into a nucleus of carbon.

tritium   a heavy isotope of hydrogen with a mass number of 3; radioactive, with a half-life of 12.3 years.

triton   the nucleus of a **tritium** atom; symbolized ^3_1H.

troposphere   the lowest layer of the atmosphere; its thickness ranges from about 11 miles at the poles to about 45 miles at the equator.

ultraviolet   the region of the electromagnetic spectrum that begins where violet light leaves off and is higher in energy and frequency than violet light.

unsaturated   an organic compound with one or more carbon-carbon double or triple bonds.

viscous   description of a liquid that flows very slowly.

volatile   description of a liquid that evaporates readily at room temperature.

volatile organic compounds   organic compounds found in the atmosphere.

vulcanization   the process by which rubber is strengthened by the addition of sulfur.

watt   a unit of power equal to one joule of energy per second.

wave equation   an equation used in quantum mechanics to describe the properties of atoms and molecules.

wavelength   the distance in meters between two adjacent crests of a wave.

X-rays   electromagnetic radiation that is higher in frequency and energy than visible or ultraviolet light.

Further Resources

The following sources offer readings related to individual nonmetal elements.

HYDROGEN
Books and Articles

Ariza, Luis Miguel. "Burning Times for Hot Fusion." *Scientific American* 282 (March 2000): 19–20. This article focuses on efforts of researchers to develop methods to produce electricity from D-T fusion.

Ashley, Steven. "Fuel Cell Phones," *Scientific American* 285 (July 2001): 25. This article describes how fuel cells can be used to power cell phones.

"Balloons: The First Aircraft to Carry Men." *Above and Beyond: The Encyclopedia of Aviation and Space Sciences,* vol. 2. Kingsport, Tenn.: New Horizons Publishers, 1968. Part of a series of books about aviation, this volume details the history of balloon flight.

Frank, Adam. "How the Big Bang Forged the First Elements," *Astronomy* (October 2007): 32. Adam Frank gives an accounting of current scientific understanding regarding nucleosynthesis during the big bang.

Heppenheimer, T. A. *A Brief History of Flight: From Balloons to Mach 3 and Beyond.* New York: John Wiley and Sons, 2001. This book provides an overview of the history of flight, including the story of the *Hindenburg.*

Hoffman, Peter. *Tomorrow's Energy: Hydrogen, Fuel Cells, and the Prospects for a Cleaner Planet.* Cambridge, Mass.: The MIT Press, 2002. Discusses the history, economics, and science of hydrogen fuel cell development.

Lemley, Brad. "Lovin' Hydrogen," *Discover,* 1 November 2001. Amory Lovins describes how a profitable, pollution-free hydrogen economy is possible.

Koppel, Tom. *Powering the Future: The Ballard Fuel Cell and the Race to Change the World.* Etobicoke, Ont.: John Wiley & Sons Canada,

1999. This book tells the story of a small Canadian company working to develop hydrogen fuel cells.

Rae, Alastair I. M. *Quantum Physics: A Beginner's Guide.* Oxford: Oneworld Publications, 2005. This book explains the wave and particle nature of atoms and their quantum behavior in a way that nonscientists can understand.

Internet Resources

Vidicom Media Productions. "Titanic of the Sky: The Hindenburg Disaster." Available online. URL: www.vidicom-tv.com/tohiburg. htm. Accessed June 30, 2008. This Web site provides a comprehensive review of the *Hindenburg* tragedy with links to a slideshow and a video.

CARBON

Books and Articles

Bruice, Paula Y. *Essential Organic Chemistry.* Upper Saddle River, N.J.: Prentice Hall, 2006. An overview of the basic chemistry of carbon and its compounds, this book assumes the reader has a background in elementary general chemistry.

Geim, A. K., and K. S. Novoselov. "The Rise of Graphene." *Nature Materials* 6 (March 2007): 183. This article describes graphene's structure, manufacture, and potential in an entirely new class of two-dimensional materials.

Libby, Willard F., and Frederick Johnson. *Radiocarbon Dating,* 2nd ed. Chicago: University of Chicago Press, 1955. Dr. Libby was awarded the Nobel Prize in Chemistry in 1960 for the development of carbon dating methods. This book is Libby's story in his own words.

Myers, Rollie J. "One-Hundred Years of pH." *Journal of Chemical Education* 87, no. 1 (2010): 30–32. A historical overview of the origin and development of the concept and measurement of pH.

Internet Resources

Jet Propulsion Laboratory, California Institute of Technology, Pasadena, CA. Available online. URL: www.spitzer.caltech.edu/Media/

happenings/20070426. Accessed on December 19, 2008. Press release from Cal Tech's Spitzer Science Center, describing the observation of polycyclic aromatic hydrocarbons in space.

NITROGEN
Books and Articles

Avenier, P., M. Taoufik, A. Lesage, X. Solans-Monfort, A. Baudouin, A. deMallmann, L. Veyre, J.-M. Basset, O. Eisenstein, L. Emsley, and E. A. Quadrelli. "Dinitrogen Dissociation on an Isolated Surface Tantalum Atom," *Science* 317 (2007): 1,056–1,060. This research article describes an alternative route to ammonia synthesis that possibly is more efficient and less expensive than the traditional Haber-Bosch process.

Leigh, G. J. *The World's Greatest Fix: A History of Nitrogen and Agriculture.* New York: Oxford, 2004. This book describes the historical understanding of agricultural methods regarding fertilizers.

Martin, Lawrence. *Scuba Diving Explained: Questions and Answers on Physiology and Medical Aspects of Scuba Diving.* Flagstaff, Ariz.: Best Publishing Co., 1997. An in-depth description of the physics of decompression sickness is given in Section G of this comprehensive book on human physiology in the ocean environment.

Schwedt, Georg. *The Essential Guide to Environmental Chemistry.* New York: John Wiley and Sons, 2001. This is a brief, elementary summary of important topics in environmental chemistry.

Internet Resources

Official Web site of the United States Environmental Protection Agency. Available online. URL: www.epa.gov/air/urbanair/nox/. Accessed on June 11, 2009. This Web page describes how nitrogen oxides affect the way we live and breathe.

PHOSPHORUS
Books and Articles

Brady, Nyle C., and Ray R. Weil. *Elements of the Nature and Properties of Soils.* Upper Saddle River, N.J.: Prentice Hall, 2003. This is an excellent resource for finding out about plant nutrients in soils.

Chemical and Engineering News 58 (2 July 2007). A production index is published annually showing the quantities of various chemicals that are manufactured in the United States and other countries.

Fleming, Graham R., and Rienk van Grondelle. "The Primary Steps of Photosynthesis." *Physics Today* 47 (February 1994): 48–55. This article gives an account of the complex, yet efficient, manner in which plants convert solar energy to chemical energy.

Turco, Richard P. *Earth Under Siege: From Air Pollution to Global Change.* New York: Oxford University Press, 2002. Describes the basic principles, physics, and effects of human-sourced air pollution.

Internet Resources

"Beautiful Pools of Pollution." Discover Magazine. Available online. URL: discovermagazine.com/photos/23-the-dark-side-of-the-green revolution/?searchterm=phosphorus. Accessed January 14, 2009. This online photo gallery shows the environmental devastation that phosphorus mining can cause.

Official Web site of the United States Environmental Protection Agency. Available online. URL: www.epa.gov. The most current environmental regulations and policies may be found here.

OXYGEN

Books and Articles

Flannagan, William A. "Energetic Oxygen Ions Stream up to Magnetosphere." *Physics Today* 30 (November 1977): 17–19. This article explains how oxygen ions are transported into the upper atmosphere and its effects there.

Jackson, Joe. *A World on Fire: A Heretic, an Aristocrat, and the Race to Discover Oxygen.* New York: Viking, 2007. This is a dramatic recounting of scientists in a competition to be the first to discover oxygen, with a backdrop steeped in the lively history of the time.

Lane, Nick. *Oxygen: The Molecule That Made the World.* Oxford, U.K.: Oxford University Press, 2002. A sweeping discussion of oxygen's role in the formation of life and its responsibility for cellular deterioration.

SULFUR
Books and Articles

Baird, Colin. *Chemistry in Your Life,* 2nd ed. New York: W. H. Freeman and Company, 2006. This is an elementary textbook in basic chemistry that describes the importance of various elements, including sulfur, in everyday life.

Kargel, Jeffrey S., Pierre Delmelle, and Douglas B. Nash. "Volcanogenic Sulfur on Earth and Io: Composition and Spectroscopy." *Icarus* 142 (November 1999): 249–280. This article compares and contrasts sulfur impurities on Io's surface with sulfur deposits on Earth.

Knoll, Andrew H. *Life on a Young Planet.* Princeton, N.J.: Princeton Science Library, 2005. A discussion of the roles of sulfur and oxygen in the evolution of life.

Kutney, Gerald. *Sulfur: History, Technology, Applications & Industry.* Norwich, N.Y.: William Andrew Publishing, 2007. This book is a scientific history of sulfur, tracking the technologies, applications, and the industry itself from ancient markets to the current global economy.

SELENIUM
Books and Articles

Hatfield, Dolph L., Marla J. Berry, and Vadim N. Gladyshev, eds. *Selenium: Its Molecular Biology and Role in Human Health.* Dordrecht, The Netherlands: Kluwer Academic Publishers, 2001. This book explains recent advances in the understanding of selenium's role in the human body.

Weiss, Herbert V., Minoru Koide, and Edward D. Goldberg. "Selenium and Sulfur in a Greenland Ice Sheet: Relation to Fossil Fuel Combustion." *Science* 172 (16 April 1971): 261–263. Describes how the ratio of selenium to sulfur in glacial ice is characteristic of terrestrial matter, and how these elements may find their way to ice sheets.

General Resources

The following sources discuss general information on the periodic table of the elements.

Books and Articles

Ball, Philip. *The Elements: A Very Short Introduction.* Oxford: Oxford University Press, 2002. This book contains useful information about the elements in general.

Chemical and Engineering News 86 (2 July 2008). A production index is published annually showing the quantities of various chemicals that are manufactured in the United States and other countries.

Considine, Douglas M., ed. *Van Nostrand's Encyclopedia of Chemistry,* 5th ed. New York: John Wiley and Sons, 2005. In addition to its coverage of traditional topics in chemistry, the encyclopedia has articles on nanotechnology, fuel cell technology, green chemistry, forensic chemistry, materials chemistry, and other areas of chemistry important to science and technology.

Cotton, F. Albert, Geoffrey Wilkinson, and Paul L. Gaus. *Basic Inorganic Chemistry,* 3rd ed. New York: John Wiley and Sons, Inc., 1995. Written for a beginning course in inorganic chemistry, this book presents information about individual elements.

Cox, P. A. *The Elements on Earth: Inorganic Chemistry in the Environment.* Oxford: Oxford University Press, 1995. There are two parts to this book—the first describes Earth and its geology and how elements and compounds are found in the environment as well as how elements are extracted from the environment, and the second part describes the sources and properties of the individual elements.

Daintith, John, ed. *The Facts On File Dictionary of Chemistry.* 4th ed. New York: Facts On File, 2005. Definitions of many of the technical terms used by chemists.

Emsley, John. *Nature's Building Blocks: An A–Z Guide to the Elements.* Oxford: Oxford University Press, 2001. Proceeding through the periodic table in alphabetical order of the elements, Emsley describes each element's important properties, biological and medical roles, and importance in history and the economy.

——. *The Elements.* Oxford: Oxford University Press, 1989. Provides a quick reference guide to the chemical, physical, nuclear, and electron shell properties of each of the elements.

Greenberg, Arthur. *Chemistry: Decade by Decade.* New York: Facts On File, 2007. An excellent book that highlights by decade the important events that occurred in chemistry during the 20th century.

Greenwood, N. N., and A. Earnshaw. *Chemistry of the Elements.* Oxford: Pergamon Press, 1984. This book is a comprehensive treatment of the chemistry of the elements.

Hall, Nina, ed. *The New Chemistry.* Cambridge: Cambridge University Press, 2000. Contains chapters devoted to the properties of metals and electrochemical energy conversion.

Hampel, Clifford A., ed. *The Encyclopedia of the Chemical Elements.* New York: Reinhold Book Corporation, 1968. In addition to articles about individual elements, this book also has articles about general topics in chemistry. Numerous authors contributed to this book, all of whom were experts in their respective fields.

Heiserman, David L. *Exploring Chemical Elements and Their Compounds.* Blue Ridge Summit, Pa.: Tab Books, 1992. This book is described by its author as "a guided tour of the periodic table for ages 12 and up," and is written at a level that is very readable for pre-college students.

Henderson, William. *Main Group Chemistry.* Cambridge: The Royal Society of Chemistry, 2002. This book is a summary of inorganic chemistry in which the elements are grouped by families.

Jolly, William L. *The Chemistry of the Non-Metals.* Englewood Cliffs, N.J.: Prentice-Hall, Inc., 1966. This book is an introduction to the chemistry of the nonmetals, including the elements covered in this book.

King, R. Bruce. *Inorganic Chemistry of Main Group Elements.* New York: Wiley-VCH, 1995. This book describes the chemistry of the elements in the s and p blocks.

Krebs, Robert E. *The History and Use of Our Earth's Chemical Elements: A Reference Guide,* 2nd ed. Westport, Conn.: Greenwood Press, 2006. Following brief introductions to the history of chemistry and atomic structure, Krebs proceeds to discuss the chemical and physical properties of the elements group (column) by group. In addition, he describes the history of each element and current uses.

Lide, David R., ed. *CRC Handbook of Chemistry and Physics,* 89th ed. Boca Raton, Fla.: CRC Press, 2008. The *CRC Handbook* has been the most authoritative, up-to-date source of scientific data for almost nine decades.

Mendeleev, Dmitri Ivanovich. *Mendeleev on the Periodic Law: Selected Writings, 1869–1905.* Mineola, N.Y.: Dover, 2005. This English translation of 13 of Mendeleev's historic articles is the first easily accessible source of his major writings.

Norman, Nicolas C. *Periodicity and the p-Block Elements.* Oxford: Oxford University Press, 1994. In addition to updating and substantially rewriting parts of the first edition, certain aspects of s-block element chemistry are now discussed explicitly.

Parker, Sybil P., ed. *McGraw-Hill Encyclopedia of Chemistry,* 2nd ed. New York: McGraw Hill, 1993. A comprehensive treatment of the chemical elements and related topics in chemistry, including expert-authored coverage of analytical chemistry, biochemistry, inorganic chemistry, physical chemistry, and polymer chemistry.

Rouvray, Dennis H., and R. Bruce King, eds. *The Periodic Table: Into the 21st Century.* Baldock, Hertfordshire, England: Research Studies Press Ltd., 2004. This is a collection of papers from the second international conference on the Periodic Table, held in memory of Harry Weiner, in Canada in July 2003. The book contains chapters on the early history and development of the periodic table, the theoretical foundations, some pedagogical aspects, the future of the table, and a brief excursion into nonelemental periodic tables.

Stwertka, Albert. *A Guide to the Elements,* 2nd ed. New York: Oxford University Press, 2002. This book explains some of the basic concepts of chemistry and traces the history and development of the periodic table of the elements in clear, nontechnical language.

Winter, Mark J., and John E. Andrew. *Foundations of Inorganic Chemistry.* Oxford: Oxford University Press, 2000. This book presents an elementary introduction to atomic structure, the periodic table, chemical bonding, oxidation and reduction, and the chemistry of the elements in the s, p, and d blocks; in addition, there is a separate chapter devoted just to the chemical and physical properties of hydrogen.

Internet Resources

American Chemical Society homepage. Available online. URL: www.chemistry.org. Accessed on December 19, 2008. Many educational resources are available.

Center for Science and Engineering Education, U.S. Department of Energy National Laboratory Operated by the University of California. Available online. URL: www.lbl.gov/Education. Accessed on June 11, 2009. Contains educational resources in biology, chemistry, physics, and astronomy.

Chemical Education Digital Library. Available online. URL: www.chemeddl.org/index.html. Accessed on December 19, 2008. Digital content intended for chemical science education. "Chemical Elements." Available online. URL: www.chemistryexplained.com/elements. Accessed December 19, 2008. Information about each of the chemical elements.

Chemical Elements.com. Available on online. URL: www.chemicalelements.com. Accessed December 19, 2008. A private Web site that originated with a school science fair project.

Chemicool. Available online. URL: www.chemicool.com. Accessed on December 19, 2008. Information about the periodic table and the chemical elements created by David D. Hsu, the Massachusetts Institute of Technology.

Journal of Chemical Education, Division of Chemical Education, American Chemical Society. Available online. URL: jchemed.chem. wisc.edu/HS/index.html. Accessed on December 19, 2008. The Web site for the premier online journal in chemical education.

Lenntech Water Treatment & Air Purification, Rotterdamseweg 402 M 2629 HH Delft, The Netherlands. Available online. URL: www. lenntech.com/Periodic-chart.htm. Accessed on December 19, 2008. Contains an interactive, printable version of the periodic table.

Los Alamos National Laboratory, Chemistry Division, Los Alamos, New Mexico. Available online. URL: periodic.lanl.gov/default.htm. Accessed on December 19, 2008. A resource on the periodic table for elementary, middle school, and high school students.

Mineral Information Institute, 8307 Shaffer Pkwy, Littleton CO 80127. Available online. URL: www.mii.org. Accessed on December 19, 2008. A large amount of information for teachers and students about rocks and minerals and the mining industry.

National Nuclear Data Center, Brookhaven National Laboratory, Upton, New York. Available online. URL: http://www.nndc.bnl.gov/ content/HistoryOfElements.html. Accessed on December 19, 2008. A worldwide resource for nuclear data.

New York Times Company, About.com: Chemistry. Available online. URL: chemistry.about.com/od/chemistryfaqs/f/element.htm. Accessed on December 19, 2008. Information about the periodic table, the elements, and chemistry in general.

Periodic Table of Comic Books, Department of Chemistry, University of Kentucky. Available online. URL: www.uky.edu/Projects/ Chemcomics. Accessed on December 19, 2008. A fun, interactive version of the periodic table.

The Periodic Table of Videos. Available online. URL: www. periodicvideos.com. Accessed on December 19, 2008. Short videos on all of the elements can be viewed. The videos can also be accessed through YouTube®.

Schmidel & Wojcik: Web Weavers. Available online. URL: quizhub. com/quiz/f-elements.cfm. Accessed on December 19, 2008. A K–12 interactive learning center that features educational quiz games for English language arts, mathematics, geography, history, earth science, biology, chemistry, and physics.

United States Geological Survey. Available online. URL: minerals.usgs. gov. Accessed on December 19, 2008. The official Web site of the Mineral Resources Program.

Web Elements, The University of Sheffield, United Kingdom. Available online. URL: www.webelements.com/index.html. Accessed on December 19, 2008. A vast amount of information about the chemical elements.

Wolfram Science. Available online. URL: demonstrations.wolfram. com/PropertiesOfChemicalElements. Accessed on December 19, 2008. Information about the chemical elements from the Wolfram Demonstration Project.

Department of Chemistry, University of Nottingham. Available online. URL: www.periodicvideos.com/. Accessed on June 12, 2009. Short videos on all of the elements can be viewed here.

Periodicals

Discover
Published by Buena Vista Magazines
114 Fifth Avenue
New York, NY 10011
Telephone: (212) 633-4400
www.discover.com
A popular monthly magazine containing easy to understand articles on a variety of scientific topics.

Journal of Chemical Education
American Chemical Society
1155 16th Street, NW
Washington, DC 20036
Telephone: (202) 872-4600
pubs.acs.org/journal/jceda8
One of the major journals for articles related to chemical education.

Nature
The Macmillan Building
4 Crinan Street
London N1 9XW
Telephone: +44 (0)20 7833 4000
www.nature.com/nature
A prestigious primary source of scientific literature.

Science
Published by the American Association for the Advancement of Science
1200 New York Avenue, NW
Washington, DC 20005
Tel: (202) 326-6417
www.sciencemag.org
One of the most highly regarded primary sources for scientific literature.

Scientific American
415 Madison Avenue
New York, NY 10017
Telephone: (212) 754-0550
www.sciam.com
A popular monthly magazine that publishes articles on a broad range of subjects and current issues in science and technology.

Index

Note: *Italic* page numbers refer to illustrations.